CONCEPTS OF MODERN PHYSICS

Manoj Bhatia

ANMOL PUBLICATIONS PVT. LTD.
NEW DELHI-110 002 (INDIA)

ANMOL PUBLICATIONS PVT. LTD.

Regd. Office: 4360/4, Ansari Road, Daryaganj,
New Delhi-110002 (India)
Tel.: 23278000, 23261597, 23286875, 23255577
Fax: 91-11-23280289
Email: anmolpub@gmail.com
Visit us at: www.anmolpublications.com

Branch Office: No. 1015, Ist Main Road, BSK IIIrd Stage
IIIrd Phase, IIIrd Block, Bangalore-560 085 (India)
Tel.: 080-41723429 • Fax: 080-26723604
Email: anmolpublicationsbangalore@gmail.com

Concepts of Modern Physics

First Edition, 2010

ISBN 978-81-261-4535-5

PRINTED IN INDIA

Printed at Mehra Offset Press, Delhi.

Contents

	Preface	*vii*
1.	Photoelectric Equation	1
2.	Theories and Hypothesis	70
3.	X-Rays and Productions	134
4.	Nuclear Instability	147
5.	Introduction to Nuclear Physics	158
6.	Radioactive Dcay and Exponential Laws	182
7.	Introduction to Rays	190
8.	Nuclear Reactors	210
9.	Conductors	228
	Index	295

Modern
presentatio
The book
modern ph
physics an
toward
applicatio
a concep
The se
historical
first to
atoms and
and follow
of atoms
atomic nu

Preface

Modern Physics is the most up-to-date, accessible presentation of physics what we examine and analyse today. The book is intended to be used in a semester course covering modern physics for students who have already had basic physics and calculus. The balance of the book leans more toward ideas than toward experimental methods and practical applications because the beginning student is better served by a conceptual framework than by a mass of details.

The sequence of topics follows a logical, rather than strictly historical, order. Relativity and quantum ideas are considered first to provide a framework for understanding the physics of atoms and nuclei. The theory of the atom is then developed, and followed by a discussion of the properties of aggregates of atoms, which includes a look at statistical mechanics. Finally atomic nuclei and elementary particles are examined.

Author

Chapter 1

Photoelectric Equation

EINSTEIN'S PHOTOELECTRIC EQUATION

To take into account the phenomenon of photoelectric emission Einstein proposed a new equation,

$hf = W + ½ mv^2$

hf = energy of incident photon

W = work function of the metal surface is the minimum amount of energy that has to be given to an electron to release it from the surface. $½ mv^2$ = the maximum kinetic energy of the emitted electrons.

If hf < W, then there is not sufficient energy to eject an electron, i.e. the threshold frequency, fo, is given by $hf_o = W$ For minimum frequency fo, the corresponding maximum wavelength λ_o is given by $h (c/l_o) = W$.

STOPPING POTENTIAL

It is that negative potential difference, Vs which just stops the electrons with maximum energy from reaching an electron collector (anode).

Two metal plates C and A are sealer in a vacuum container. Light of appropriate frequency falls on C which is called cathode or emitter. Electrons are emitted by C and collected by A called anode or collector. The potential difference between C and A can be changed with the help of batteries, rheostat and commutator.

The anode potential can be made +ve or –ve w.r.t. the cathode. The electrons collected by A flow through the ammeter, batteries etc. and are back to C; an electric current is established in the circuit. Such a current is called a photocurrent.

When the anode, A is given a positive potential w.r.t. cathode, electrons are attracted towards anode and the photocurrent increases. If the potential of A is gradually increased, photocurrent also increase and finally becomes constant and is known as saturation current.

If the potential of the anode is made negative w.r.t. cathode, C, then the electrons are repelled by A and the photocurrent decreases. At a certary value of this negative potential, Vs the photocurrent is totally stopped. The smallest value of the anode potential which just stops the photocurrent is called stopping potential.

The stopping potential is related to maximum kinetic energy of the ejected electrons.

The stopping potential, Vs is:

- Proportional to maximum kinetic energy of ejected electron
- Proportional to the frequency of radiation and
- Independent of intensity of radiation

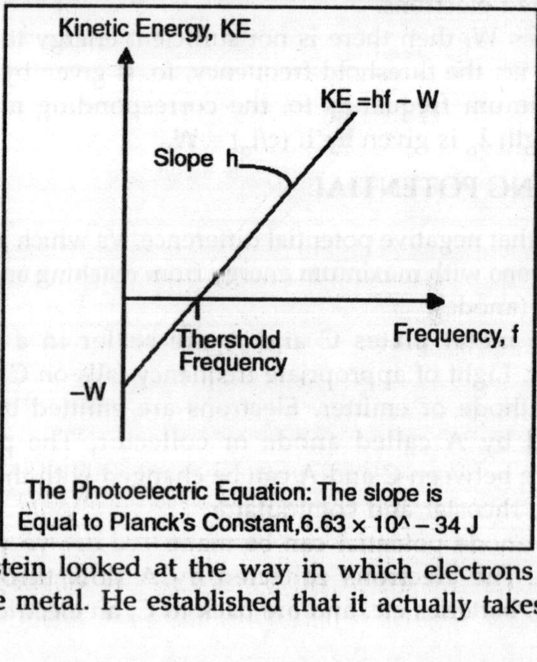

The Photoelectric Equation: The slope is Equal to Planck's Constant,6.63 × 10^ – 34 J

Einstein looked at the way in which electrons tried to leave the metal. He established that it actually takes energy for an electron to leave the surface of the metal. He called this energy the work function, Φ.

He then applied the Conservation of Energy Principle for an electron receiving energy and leaving the metal. Compare energy received and used by the electron at the surface. Energy supplied to the electron = hf (energy from the photon)

Energy used by the electron is either:

Used as work function, Φ, to escape from the surface,

or

is left with the electron after it has escaped, in which case it is in the form of E_k.

So for each electron at the surface which leaves the metal:

Energy supplied = energy used

or

$$hf = F + E_{kmax}$$

The E_{kmax} is there because only those electrons on the surface will have the maximum kinetic energy on leaving.

Electrons from deep inside which make it to the surface and still have enough energy to escape will have used some energy getting to the surface. These electrons will therefore have less energy left once they are free of the metal and so they will have E_k less than the maximum possible.

Notice that if you consider the special case where the energy of the arriving photons is hf_o(i.e. the energy of the threshold frequency photons), even the electrons on the surface will only have enough energy to overcome the work function and no more.

So once they have escaped they will have no energy left. Their E_k = zero.

In this case, the photoelectric equation becomes:

$$hf_0 = \Phi$$

This is a useful way of calculating the work function for a material. *Also it allows you to rewrite the Photoelectric equation as follows*:

$$hf = hf_0 + E_{k\,max}$$

or

$$h(f - f_0) = \frac{1}{2}mv^2_{max}$$

THE PHOTOELECTRIC EFFECT

In 1887, the German physicist Heinrich Rudolf Hertz discovered an interesting property of matter. This property is that physical materials emit charged particles when they absorb radiant energy (eg, light).

Of course, not all substances absorb radiant energy, and the ones that don't will not emit charged particles. But it could readily be established that some substances do behave this way.

Hertz initially observed that the minimum voltage required to draw sparks from a pair of metallic electrodes was reduced when they were bathed in ultaviolet (UV) light, such as from a mercury vapour lamp. The more intense the UV light, the lower the required voltage became.

In the broadest sense, the substance in question can be solid, liquid, or gaseous; the radiant energy must be visible light, UV; X-rays; or gamma rays (cosmic radiation). The charged particles can be electrons or ions. However, in general practical use, the substance is a metal plate and the charged particles are electrons.

In any case, a second German physicist, Philipp Lenard, studied this phenomenon using a metal plate, and in 1900 concluded that the charged particles emitted were the same as those found in cathode rays. That is, they were electrons. By 1902, it had been shown that the resulting current (called photoelectric current because it was caused by light) is proportional to the intensity of the light causing it for any given frequency of light energy, and that the maximum kinetic energy imparted to any electron is independent of the intensity of the light, but is directly proportional to the frequency of the light.

Albert Einstein worked out the primary equation involved, and by 1912, the requisite measurements could be made with high precision. These experiments confirmed the conclusion that light is not a continuously wave-like phenomenon, but rather involves particle-like"corpuscles" of energy, now called photons, which are the quanta of electromagnetic energy.

THE EXPERIMENT

The experiment itself is now performed in appropriate physics classes at most colleges and universities, and possibly in some high schools. It isn't difficult to perform, and if proper care is used, will give quite accurate results.

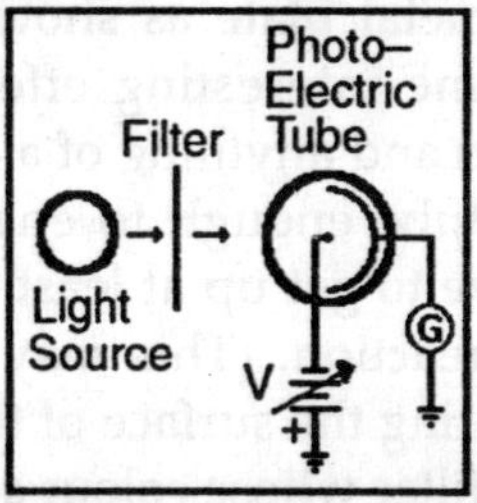

We begin with a photoelectric tube (or phototube), which is a vacuum tube containing a metal plate curved into a half-cylinder (the anode) and a thin wire electrode (the cathode) along the axis of the cylinder. The figure to the right shows this as seen from the top.

We use a very sensitive metre, called a galvenometre (G), to measure the current passing through the tube, and a variable voltage source (V) to control the voltage applied between the two electrodes inside the tube. Finally, we add a light source (usually a mercury vapour lamp, but it can be an arc light or even a tungsten filament light bulb), and a coloured filter to limit the light striking the photoelectric tube to a single frequency.

With no light, of course, the current through the tube will be zero, regardless of the applied voltage. (Yes, there is a slight capacitance between the metal plate and the wire, and a high enough voltage will cause an arc. But the capacitance is very slight and will charge quickly, and we won't be using anywhere near enough voltage to cause any problems.) So we turn on the light, place one of our coloured filters between the light and the phototube, and begin our measurements.

The first thing we note is that if we reverse the polarity of the voltage source V, increasing the voltage or the intensity of the light will increase the current flow. This seems logical;

the light is"kicking" electrons off the metal plate, and they are attracted to the now positively-charged wire electrode. The more light, the more electrons and hence the higher the current.

However, when we make the wire electrode negative with respect to the metal plate as shown in this figure, we begin to observe some interesting effects. First, red light, infrared (IR) radiation and anything of a lower frequency will not excite the phototube enough to enable current to flow. You pretty much have to get up at least to green light before you can measure a reaction. (This may vary depending on the specific metal coating the surface of the curved electrode.)

Once you find a filter whose colour allows current to flow through the tube, you can begin to take your measurements.

PLOTTING THE RESULTS

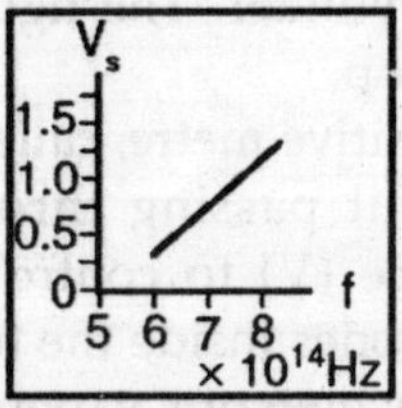

For the experiment itself, the object is to find some voltage, V, which will just prevent any current flow through the phototube.

This is the *stopping voltage,* V_s which is just barely enough to prevent any electrons emitted from the anode plate from reaching the cathode wire. Plot this point on a graph showing voltage on the Y axis and the frequency of the filtered light on the X axis. Thus, you are generating a graph of stopping voltage (V_s) as a function of the frequency of the light reaching the phototube.

Note: The graph shown to the right does not necessarily match the results you would obtain from this experiment, although your results should be similar. The specific materials used in the phototube will have a significant effect on your measured and plotted results.

One interesting fact here is that once you have set the

applied voltage to V_s, increasing the intensity of the light reaching the phototube (but not changing the filter) will not enable current to flow again.

This is a key point. It demonstrates that each photon of a given frequency has exactly the same energy as every other photon of that frequency. Higher intensity means that more photons reach the target per second for more total energy delivered, but each photon still has the same energy as before.

Next, we change the filter to some other colour and as before seek to find V_s for this particular frequency of light. Again, we plot this point on our graph. We continue this for every filter we have, and then fill in the spaces between our plotted points. If we've been careful enough with our measurements, we will find that the graph plots one of two ways.

For all frequencies above a certain minimum, we have a straight line showing a linear relationship between V_s and frequency. Below that minimum frequency, there is no current flow at all, regardless of the applied voltage. If you did not use any filters for these frequencies, this cutoff frequency will not appear directly on your graph. However, you can find this point from your graph. The frequency at which $V_s = 0$ is the limit to which you can take this experiment; lower frequencies of light will not affect the phototube.

FACTORS

There are two basic factors that control the results of this experiment. First, it takes a certain amount of energy to free an electron from the surface of the metal anode of the phototube.

This is known as the *work factor*. Second, the arriving light clearly imparts a certain kinetic energy to the freed electrons, which enables them to overcome the repulsion effect of the more negative cathode wire, and still cause a current to flow through the phototube. It is that kinetic energy that we are measuring when we adjust the applied voltage V to determine the precise value of V_s. Let's look at these two factors separately.

The Work Factor

When light energy reaches the metal anode, it may be sufficient to push an electron out of its orbit around its parent atom, and move it towards the surface. If this electron is from an atom at the surface of the anode, a certain amount of energy is still required to kick the electron away from the metal and let it fly through space. This energy is a measure of the amount of work required to remove the electrum from its initial environment. Therefore, it is known as the *work factor* of that particular metal.

Different metals have different work factors, but the work factor of any metal is a characteristic of the metal itself, and does not change for different frequencies of light. Thus, phototubes made of different metal materials will have different work factors, but the work factor for any given phototube will be constant in the experiment.

If the arriving light energy bypasses the first layer or two of atoms at the surface and then frees an electron, that electron will lose energy as it travels past other atoms before it is able to leave the surface. This electron will have less kinetic energy traveling through the vacuum of the tube than one released from a surface atom.

The work factor varies from about 2.2 electron volts (eV) for lithium to 6.35 eV for platinum. Any frequency of light which cannot impart enough energy to the electrons in these metals to overcome the work factor will fail to cause a current to flow through the experimental circuit.

Measuring the Kinetic Energy of the Electrons

Once an electron has been freed from the surface, it is still moving, and therefore has a certain amount of kinetic energy. We can determine the kinetic energy of the most energetic electrons (often designated k_{max}) by determining the voltage needed to just prevent them from reaching the cathode of the phototube. The energy involved is then the product of the stopping voltage, V_s, and the electric charge on the electron. That charge is a fixed, known value, designated by the letter "e.". Thus, the product of the charge *e* and applied voltage *V* is designated *eV* and represents energy measured in electron-volts.

So long as we keep our energy units in eV, our experimental stopping voltage, V_s, is a direct measure of the energy needed to exactly cancel the kinetic energy of the free electrons, and hence of the energy imparted to those electrons by the arriving light.

Interpreting the Experimental Results

Although the exact measured results you get will depend on the exact material coating the anode of your phototube, your graph should turn out to be a straight line sloped to show that Vs increases with the frequency of the light used for that measurement.

The general equation for this type of line is:

$$Y = mX + B$$

In this equation, m is the numerical slope of the line, and B is the Y-intercept. In this particular experiment, the general equation becomes:

$$k_{max} = eV_s = hf - W$$

In this expression,

- k_{max} is the maximum kinetic energy of any electron freed from the anode;
- e is the charge on any single electron;
- V_s is the stopping voltage at which conduction is just barely prevented;
- h is a constant factor;
- f is the frequency of light used in one particular step in the experiment;
- W is the work factor of the specific material coating the anode of the phototube.

Even if your filters do not include the specific colour whose frequency exactly matches the work function of your phototube's anode, you can extrapolate this value by extending your graph to the frequency (X) axis, which corresponds to $V_s = 0$. At this frequency, hf = W. Any light at a higher frequency can be used in this experiment. Light at a lower frequency cannot free electrons from the anode of the phototube at all; there is insufficient energy in such light, regardless of its intensity.

This inherent cutoff frequency varies according to the work factor of the phototube's anode, but always works the same way. As a consequence, it is not possible that light (or any electromagnetic wave) can have a continuously variable energy content. Rather, it must be made up of small"corpuscles" of energy such that each"corpuscle" either does or does not have enough internal energy to free one electron from the surface of the anode. More intense light at that frequency has more"corpuscles", but they all have the same amount of energy. If this last point were not true, the higher-energy"corpuscles" could still free electrons that lower-energy"corpuscles" could not affect.

These"corpuscles" are the quanta of electromagnetic energy, and are named photons. The remaining factor that can be determined experimentally from the graphed results of this experiment is the slope of the line. This value turns out to be just about 4.135×10^{-15} eV·seconds - which is Planck's Constant. Since the freed electrons had a maximum kinetic energy of hf - W electron-volts, the energy in the photon that kicked the electron loose must have been E = hf. This expression relates the energy of the photon to its frequency throughout the electromagnetic spectrum.

The inescapable conclusion of this experiment is that light (or any electromagnetic radiation) is not a continuous phenomenon as originally believed, but rather is made up of discrete"bundles" of energy, now called photons, each of which exists independently of all other photons. It can give up its energy to a physical object such as an electron, but does not transfer energy to or from other photons or manifestations of energy. The energy content of the photon is proportional to its frequency (related by Planck's Constant), but is not controlled by anything else.

EXPERIMENT THE PHOTOELECTRIC EFFECT

The photoelectric effect is important in that it provides one of the strongest experimental demonstrations of the quantum nature of light. When light is shone upon a metal surface, it can liberate electrons from the surface to produce a photoelectric current. The photoelectric current depends on the intensity and wavelength of the incident light.

Apparatus

- Digital voltmeter
- H/e apparatus box with support base assembly and coupling bar, batteries included
- Lens grating assembly
- Mercury vapour light source
- Light aperture assembly and light block for mercury vapour light source
- Connecting cords
- Selective wavelength filters (green or yellow) for various mercury lamp wavelength
- Variable transmission filter to vary light intensity

Theory

All attempts to explain the photoelectric effect using the wave theory of light were unsuccessful and it was Albert Einstein, in 1905, who correctly interpreted the observations. Making use of Planck's quantum ideas, Einstein postulated that light consists of a stream of discrete bundles of energy, called photons.

Each photon possesses an energy E given by,

$$E = hf \qquad (1)$$

where,

h is Planck's constant

f is the frequency of the light.

Following this assumption, the photoelectric effect is then the result of each photon transferring its entire energy to an electron in the metal. Some of the energy is used to tear the electron loose from the metal, while any excess is used to give kinetic energy (motion) to the ejected electron. This leads to Einstein's photoelectric equation for which he won the Nobel Prize namely:

$$hf = W + 1/2\ mv \qquad (2)$$

The quantity W is commonly called the"work function" and depends on the type of metal used. For any given metal, the value of W in equation (2) is the minimum energy needed to liberate the electrons, and 1/2 mv2 represents the kinetic energy of the ejected electron.

In equation (2), the photon energy hf must exceed the work function W in order for electrons to be ejected from the metal and possess kinetic energy.

If the frequency f is less than the threshold frequency f., where hf. = W, no photoelectrons are produced. In this experiment, light of a known frequency is shone upon the alkali metallic surface (potassium) of an evacuated tube or photocell. Electrons are r ejected from the cathode and some reach the collector producing a photoelectric current.

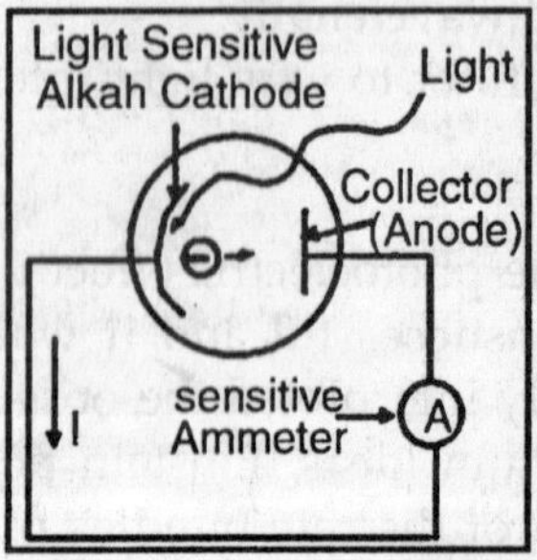

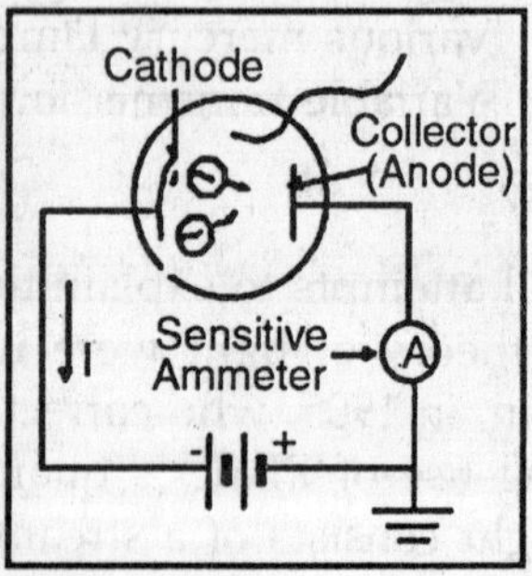

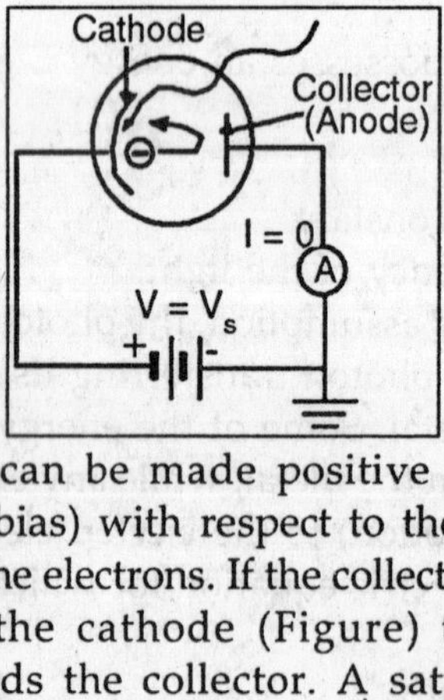

The cathode can be made positive (forward bias) or negative (reverse bias) with respect to the collector (anode) to attract or repel the electrons. If the collector is made positive with respect to the cathode (Figure) the electrons are accelerated towards the collector. A saturation current is observed as V is increased, as shown Figur. The level of the saturation current is proportional to the intensity of the incident light. When the collector is negative with respect to the cathode (Figure), the electrons are repelled from the anode.

Then only electrons leaving the emitter surface with initial kinetic energy greater than the potential energy between the cathode and collector will reach the anode and register a current. The potential difference can be increased until no electrons reach the collector and the current is stopped. This potential is called the stopping potential V_S, and is given by equation (3).

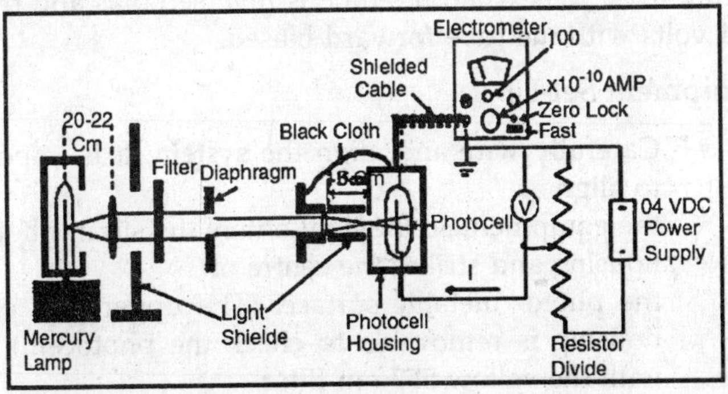

$$KE_{max} = e\,|V_S| = hf - W \quad (3)$$

This equation can be tested experimentally by measuring the stopping voltage VS for various frequencies of light f. The value of h and W can then be found from the graph of stopping voltage v_s light frequency.

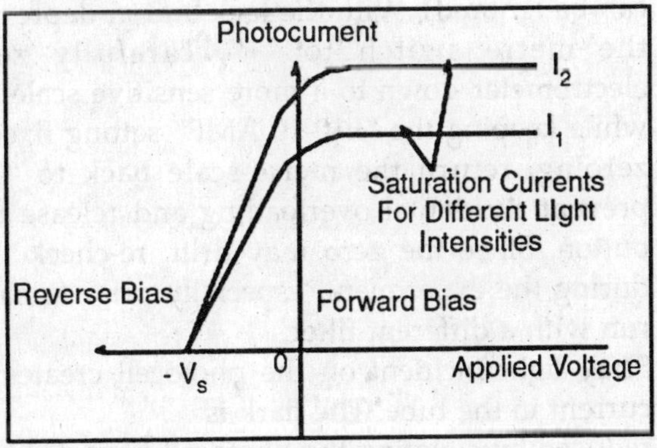

Experimental Design

The arrangement of the apparatus and the wiring.

In this diagram the tube is reversed biased. Light from a mercury lamp which passes through an interference filter associated with one of the mercury spectral lines is used to illuminate the photocell.

A very sensitive metre, called an electrometer, is used to measure the photoelectric current I. The voltage to the photocell is provided by means of a DC power supply and a resistive divider which can be adjusted to vary the voltage V from 0 to -4 volts when the tube is reversed bias and from 0 to 4 volts with the tube forward biased.

Equipment Set-Up

- Carefully wire and align the system. It is important to align the equipment so the light enters the slit in photocell housing and strikes the centre of the photo- metallic surface. (The upper part of the housing is removable to check the photocell.) Start with the yellow 577 nm filter.
- Set the electrometer to the "x10-10 AMP" setting and the metre-scale to "100". Set the NORMAL/FAST switch to FAST. (On FAST, the electrometer input resistance is negligible so that the IR drop across the electrometer can be ignored). With the lock button depressed and the metre switch to "+", carefully zero the electrometer down to a more sensitive scale (say.01) while keeping the "x10-10 AMP" setting fixed. After zeroing, return the metre-scale back to "100" to prevent accidental overloading and release the lock button. Since the zero may drift, re-check the zero during the experiment, especially prior to each new run with a different filter.
- Stray light incident on the photocell creates a dark current in the tube. The dark is reduced by covering the photocell housing.

- The stopping voltage is measured using six different frequencies of light from the mercury spectrum. The individual wavelengths are selected using the interference filters listed below.
- The experiment works best if the saturation currents are the same intensity for each spectral line. To obtain the saturation current the tube must be forward biased. The tube is reversed biased. To forward bias the tube reverse the connections to the power supply. Since yellow is a weak spectral line, the saturation current for this line with the diaphragm wide open is used for all but the very weak blue-green line.

Filter	**λ±10 (nm)**	**f (× 10^{14}Hz)**
Yellow	577.0	5.196
Green	546.1	5.490
Blue-green	491.6	6.098
Blue	433.9	6.909
Violet	404.7	7.408
Ultra-violet	365.0	8.213

Procedure

- Starting with the yellow filter forward bias the tube to 4 volts with the diaphragm wide open. Adjust the optics for maximum current. Record the saturation current level.
- Reverse bias the tube and measure the stopping voltage VS when the current I=0.
- Place the green filter into the holder. Forward bias the tube with 4 volts. Set the saturation current level to the yellow filter value by adjusting the diaphragm opening. Reverse bias the tube and measure the stopping voltage VS when the current I=0.
- Repeat step 3 using the Blue, Violet and Ultra-violet filters
- Since the Blue-green filter is extremely weak

determine the stopping voltage with diaphragm wide open.

- To investigate the affect of the light intensity on the stopping voltage, obtain measurements of the photoelectric current I as a function of retarding. Start with the tube forward biased at 4 volts and decrease the voltage until the stopping voltage is reached. Use the blue filter and start with the diaphragm wide open. It will be necessary to reverse bias the tube when V = 0. Repeat the measurements using two smaller diaphragm settings.

MATTER WAVES

We have seen that in certain experiments electrons produce patterns attributed to waves. From this observation, we concluded that electrons have wave-like properties. We also concluded that as the energy of electrons increases their wavelength decreases. Electrons are a form of matter, so these waves are called matter waves. In this chapter we will relate quantitative features of matter waves with familiar, measurable physical quantities such as energy, mass, and momentum. We will apply these relationships to forms of matter other than electrons, and see how these results can be applied to the electron microscope.

Matter waves, also called de Broglie waves, are the wave-like nature of all matter, including the atoms that make up your body. One of the first and most important findings of quantum physics is that electrons have a dual wave particle nature. It soon became apparent that all matter has this dual nature, but because conventional matter has a high momentum relative to electrons, the wavelength of the matter waves is very small, and in most cases barely noticeable. For instance, the wavelength of the matter that makes up a person is on the order of 10^{-35} metres, far smaller than can be observed using current measurement technologies.

The concept of matter waves was first elucidated by French physicist Louis de Broglie, who was extending early theories proposed by Albert Einstein, Max Planck, and Neils

Bohr. Bohr primarily studied the quantum behaviour of hydrogen atoms, whereas de Broglie tried to extend these ideas to determine a wavelength equation for all matter. De Broglie came up with a theory and presented it in his 1924 PhD thesis, for which he was awarded the Nobel Prize for Physics. This was the first case where the Nobel Prize was awarded for a PhD thesis.

Equations known as the de Broglie relations describe the dual wave-particle nature of all matter. These relations state that the wavelength of a particle is inversely proportional to its momentum (mass times velocity) and its frequency is proportional to its kinetic energy, which is a frame-dependent (relative) value. Thus, particles with low momentum, such as electrons at room temperature, have a de Broglie wavelength of about 8 nanometers. Particles with even lower momentum, such as helium atoms at temperatures of only a few nanokelvins, might have matter waves with wavelengths as long as a few microns. Under such unusual conditions, the realities of the quantum world are almost brought up into a the macroscale realm.

De Broglie's theories of matter waves were confirmed in 1927, when Bell Labs scientists Lester Germer and Clinton Davisson fired slow-moving electrons at a crystalline nickel target. The resulting diffraction pattern demonstrated the wave-like characteristics of the electrons, similar to those known to be displayed by photons such as those inx-rays. The matter waves were only able to be observed in this instance because the electrons used to produce them had very low momentum. Since 1927, the wave-like nature of various other elementary particles has been demonstrated empirically.

DEBROGLIE WAVELENGTH

Louis de Broglie was the first person to establish an equation for the relationship between an electron's momentum and its wavelength. He concluded that:

$$\text{Wavelength} = \frac{\text{Planck's Consant}}{\text{momentum}} \quad \lambda = \frac{h}{p}$$

where h is a number called Planck's constant (named after Max Planck) and is equal to 6.63×10^{-34} J × s or 4.14×10^{-15} eV × s.

When we observe electron diffraction, the electrons' kinetic energy is easier to measure than their momentum, so we write the de Broglie wavelength as,

$$\text{Wavelength} = \sqrt{\frac{(\text{Planck's Consant})^2}{2 \times \text{mass} \times \text{kinetic energy}}}$$

or

$$\lambda = \sqrt{\frac{h^2}{2mKe}}$$

This equation is consistent with our results in the previous activity - as the energy increases the wavelength decreases.

Changing Masses, Changing Patterns

While we have learned about this equation using electrons, it can be used for any type of matter. So, you might wonder why we do not see wave effects for large objects. To begin to understand and experiment with objects with masses greater than that of an electron, run the simulation of the experiment for electrons. Throughout this experiment keep the same energy.

DOUBLE SLIT DIFFRACTION

Choosing the Particle Source

The particle sources (white cylinders) pointing away from you on the centre left of the screen are not labeled, but if you move your cursor over them, the label appears. As soon as you spot the source whose spectra you wish to observe, click on it. The existing source that is in the holder will retreat, and the chosen source appears ready to fire in the holder. You may choose from electrons, protons, neutrons, photons, or pions.

Changing Wavelength of the Particles

Drag the slider below the array of sources to change the energy of the chosen particle. If your chosen particle is a photon, you will be presented with a slider that allows you to control the wavelength of the photon rather than its energy.

Changing the Slit Separation

Drag the slider at the bottom right of the screen to change the slit separation. You will see the slits come together or go apart in the experimental view, as you drag the slider.

Changing the Particles Per Second

Control the rate at which particles are emitted by dragging the ?Particles per second? slider that is just above the ?Slit Separation? slider.

You can decrease the particle rate to as low as 1 particle per second. This feature is especially useful if you are interested in predicting where the particles are likely to appear on the screen and verify your predictions with the programme.

Observing the Pattern Develop

After you have selected the particle, adjusted its energy, changed the slit separation, and controlled the number of particles that will be emitted by the source per second, you are ready to start the simulation.

Click the"Start" button that is directly below the array of sources. You will observe the diffraction pattern develop, and a counter for the total number of particles on the screen being incremented. The"Start" button changes into the"Stop" button, that can be clicked at any time to stop the simulation.

Comparing Diffraction Patterns

Once a diffraction pattern has been created it may be dragged (with the left mouse held down) to the right where it will be saved. In addition to the pattern on the screen the saved pattern will include information about the type of particle, energy or wavelength, slit width, and number of particles. Up to 4 diffraction patterns may be saved and compared at any one time.

CHANGING SLIT SEPARATIONS, CHANGING PATTERNS

Now, investigate how the patterns change as the separation between slits changes. Pick one particle and one energy; change the slit separation to answer the following. How does the pattern change as the slit separation increases? For patterns to be formed, the separation between the slits must be comparable to the wavelength of the waves passing through them. Thus, as your separation became very large

VERY LARGE MASSES

While wave behaviour is exhibited by electrons, pions, neutrons and protons; we do not observe similar behaviour for large objects such as gnats or humans. As an example, we will consider why diffraction doesn't cause a gnat to look like several gnats as it flies through window blinds. Start with a gnat whose mass is.001 kg, and speed is 0.10 m/s.

Approximately what would the spacing between the window blinds have to be for the gnat to create a pattern as it flew through? Why? Is it necessary for a gnat to worry about creating a pattern as it flies through the blinds? Why? How would a human being's de Broglie wavelength compare to that of a gnat? Why?

A SIMPLIFICATION FOR ELECTRONS

Because we know the values of Planck's Constant and the electron's mass, we can use them to simplify the equation to apply only to electrons as

$$\text{Wavelength} = \sqrt{\frac{1.5\text{nm}^2 \cdot \text{eV}}{\text{kinetic energy}}}$$

or

$$\lambda = \sqrt{\frac{1.5\text{nm}^2 \cdot \text{eV}}{\text{KE}}}$$

[ELECTRONS ONLY]

where the kinetic energy has been measured in electron volts (eV). The result, l, is the electron's wavelength in units of nanometers. This equation works only for electrons, so use it carefully.

How does the relationship between energy and wavelength in this formula compare with the relationship that you observed with simulated electron interference?

WAVING WAVE

A valid question to ask is:"What is waving with these matter waves?" Unfortunately, the answer is not an easy one. We never observe a matter wave directly; we only see results that can be explained by them. The matter wave is an abstraction that allows us to explain observations. In the we will look at what information is contained in waves of matter. We will examine the features of matter waves that describe simple properties such as the location of an object.

APPLICATION OF MATTER WAVES

Because electrons behave as waves, they can be used to"illuminate" objects in a manner similar to light. An electron microscope is an instrument that takes advantage of this situation. Electrons are given energy by accelerating them in a manner similar to the way a TV tube works. Then, using magnetic fields, they are directed at an object of interest. The electrons are focused to illuminate the object, and then to form the image of that object. A schematic diagram is shown in the diagram. This system can be used to look closely at very small objects.

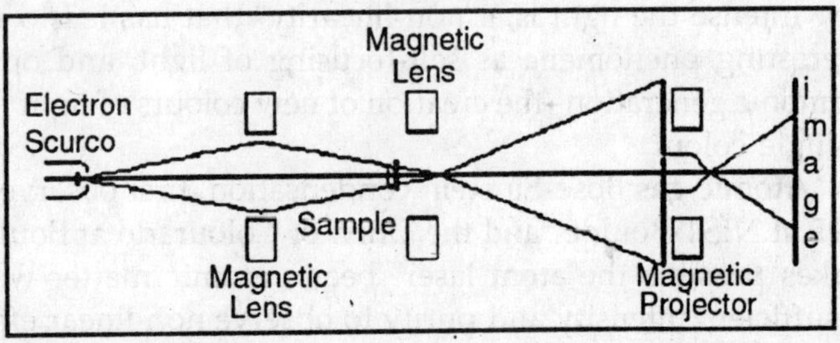

The wavelengths of the electrons are related to their

kinetic energies. In electron microscopes, wavelengths as much as 100000 times smaller than those of visible light can be achieved. With such small wavelengths, electron microscopes can reveal features that are as small as 0.000000001 metres (1 nm).

Electron Interference and Star Trek

A situation where matter waves could become important is the Star Trek transporter. We are not sure how a transporter would really work, but for the purposes of this activity, let us suppose that it decomposes a person into his or her component atoms. Then, it sends the atoms to a new location where the person is reconstructed.

Consider transporting Captain Janeway of the Voyager by such a method. She wishes to reach her new location quickly, so her atoms are sent out of the ship at 10% of the speed of light (3×10^7 m/s). Assume that her atoms have a mass of 10^{-26} kg.

Each atom must be transmitted through the titanium hull of the starship. The titanium can be considered as a large number of slits separated by 1 nm. Must the designers of the transporter be concerned about diffraction effects as the captain's atoms are beamed through the hull of the Enterprise?

MATTER WAVES

The invention of the laser ushered in a new field of experimental study in the early 1960s--non-linear optics. Because laser light can be so intense, it can actually change the index of refraction of a material through which it passes. Thus, the behaviour of a transparent medium can depend on how intense the light is, a"non-linearity" that has led to such interesting phenomena as self-focusing of light, and optical harmonic generation--the creation of new colours of light from a single colour.

Atomic-gas Bose-Einstein condensation, first observed in 1995 at NIST-Boulder and the Univ. of Colourado at Boulder, makes possible the"atom laser" beam, atomic matter waves of sufficient intensity and purity to observe non-linear effects in"atom optics"--the atomic analog to ordinary or"photon" optics.

A key difference is that while non-linear photon optics requires a non-linear medium through which the light can propagate, the non-linearity is automatically present in atom optics because of atom-atom collisional interactions, something that is virtually absent for photons in vacuum.

One important non-linear optics phenomenon is four-wave-mixing (4WM), in which three optical fields mix in a non-linear medium and create a fourth wave. The frequency and momentum of the fourth wave's photons is equal to a sum or difference among the three input frequencies and momenta.

In experiments at the National Institute of Standards and Technology in Gaithersburg, Maryland, we have performed four wave mixing of atomic matter waves, the first non-linear optics with de Broglie waves. From a single Bose-Einstein condensate we create three coherent matter waves with different velocities. These three waves mix with one another because of their collisional interactions (intrinsic non-linearity) and produce a fourth wave. We conserve the total number of atoms because for each atom appearing in the new, fourth wave, an atom disappears from each of two other waves and one atom is added to the remaining wave.

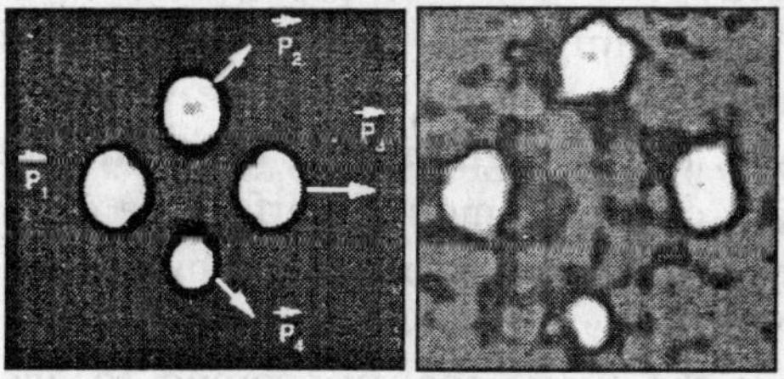

The process of matter wave 4WM can be understood as stimulated elastic scattering (analogous to stimulated emission of radiation, the basis for the optical laser) in which two atoms collide in the presence of a strong atomic matter wave, and one of the atoms is stimulated to scatter into the already-present matter wave, amplifying it. The other atom must scatter into the previously non-existent fourth wave because of energy and momentum conservation. This fourth wave is the phase conjugate of the wave that is amplified.

An alternative way of explaining atomic 4WM is that two of the original matter waves create a standing wave, a spatial corrugation of the atomic density. The third de Broglie wave Bragg-reflects from this grating, creating the fourth wave. Satisfying the condition for Bragg reflection is equivalent to satisfying energy and momentum conservation.

One of the fundamental tenets of quantum mechanics is the concept of wave-particle duality: sometimes things act like particles, and sometimes like waves. This work now extends the wave-like character of atoms into the non-linear domain. Non-linearity has been key in the development of quantum optics, which explores the uniquely quantum mechanical aspects of light. We can now look forward to the analogous development of quantum atom optics.

LOUIS DE BROGLIE AND MATTER WAVES

During the last three and a half centuries scientists have spent a great deal of time discussing and investigating the nature of light. In the 17th century Isaac Newton believed that light rays consisted of streams of very small particles. This corpuscular theory persisted for many years though Christian Huygens, a contemporary of Newton, had the notion that light might be transmitted by vibrations (i.e., waves) in the ether.

However, at the beginning of the 19th century Thomas Youngcarried out his famous interference experiments. His observations could be explained only by assuming that light is transmitted as waves and not as a stream of particles. Furthermore the wave theory seemed to account for all experimental observations made at that time, and it appeared that this theory had replaced the corpuscular theory for all time.

Then, at the end of the 19th century it was found that under certain conditions electrons were liberated when light fell on a surface. The wave theory could not explain this photoelectric effect. This startling new discovery left physicists with a serious dilemma. The photoelectric effect could best be explained by reverting to the corpuscular theory, although almost all other evidence pointed to light's being a form of wave motion.

These were some of the theoretical problems which faced physicists when Louis de Broglie, a young Frenchman of noble birth, came on the scene. In a thesis published in 1922, when he was 30 years old, he suggested that light could behave either as a wave or as a stream of particles, but not both at the same time.

He argued that if light which was normally a wave motion could take on a corpuscular (particle) form, then small particles such as electrons could also have wavelike characteristics associated with them. However, he had to wait five years for the evidence. Telephone Laboratory succeeded in diffracting (bending into shadows) a beam of electrons using a crystal as a diffraction grating.

De Broglie's dual theory can be applied to any moving particle whatever its nature. The wavelength (λ)of the"de Broglie" wave (the wave associated with the particle) is found by dividing Planck's constant (h)by the momentum of the particle (p).

Thus:

$$\lambda = h/p = h/mv$$

where m is the mass of the particle and v its velocity. Wavelengths calculated from this equation are often called de Broglie wavelengths.

Louis Victor de Broglie was born at Dieppe in France in 1892. His elder brother, Maurice, the sixth Duc de Broglie, was also a physicist of some note.

In recognition of his contribution to the advance of theoretical physics Louis de Broglie was awarded the 1929 Nobel Prize in Physics. He served as professor of theoretical physics at the University of Paris where he had previously received his training.

QUANTIZATION

Quantization refers to the process of approximating the

continuous set of values in the image data with a finite (preferably small) set of values. The input to a quantizer is the original data, and the output is always one among a finite number of levels. The quantizer is a function whose set of output values are discrete, and usually finite. Obviously, this is a process of approximation, and a good quantizer is one which represents the original signal with minimum loss or distortion.

There are two types of quantization - Scalar Quantization and Vector Quantization. In scalar quantization, each input symbol is treated separately in producing the output, while in vector quantization the input symbols are clubbed together in groups called vectors, and processed to give the output. This clubbing of data and treating them as a single unit increases the optimality of the vector quantizer, but at the cost of increased computational complexity. Here, we'll take a look at scalar quantization.

A quantizer can be specified by its input partitions and output levels (also called reproduction points). If the input range is divided into levels of equal spacing, then the quantizer is termed as a Uniform Quantizer, and if not, it is termed as a Non-Uniform Quantizer.

A uniform quantizer can be easily specified by its lower bound and the step size. Also, implementing a uniform quantizer is easier than a non-uniform quantizer. Take a look at the uniform quantizer shown below. If the input falls between n*r and (n+1)*r, the quantizer outputs the symbol n.

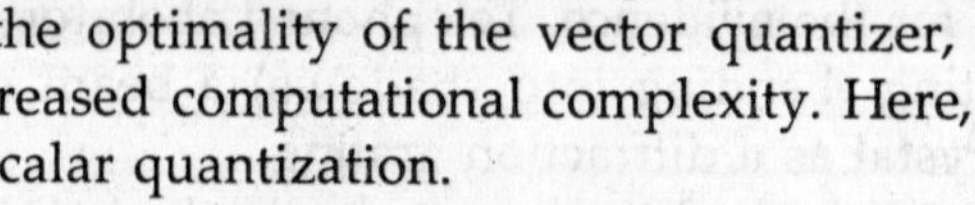

Fig. A uniform Quantizer

Just the same way a quantizer partitions its input and outputs discrete levels, a Dequantizer is one which receives the output levels of a quantizer and converts them into normal data, by translating each level into a'reproduction point' in the actual range of data. It can be seen from literature, that the optimum quantizer (encoder) and optimum dequantizer (decoder) must satisfy the following conditions.

- Given the output levels or partitions of the encoder, the best decoder is one that puts the reproduction points x' on the centres of mass of the partitions. This is known as centroid condition.
- Given the reproduction points of the decoder, the best encoder is one that puts the partition boundaries exactly in the middle of the reproduction points, i.e. each x is translated to its nearest reproduction point. This is known as nearest neighbour condition.

The quantization error (x - x') is used as a measure of the optimality of the quantizer and dequantizer.

OCTREE COLOUR QUANTIZATION

Octrees

Octrees are a very efficient data structure to store 3 dimensional data of any form. Their big advantage is, that you can search and insert in such an octree very fast. However, deleting and moving entries is a slow proccess. If you're familiar with binary trees you'll have no problem to understand the basic concept of the octrees. They're just like binary trees but have 8 subnodes instead of two. However, a short refesh of trees can't hurt, so you might read this chapter to refresh your knowledge. If all this is new to you you should read the entire article, even if you don't understand it. After you saw an example how to use them you'll surely understand them.

A Octree is a tree-structure which contains data. The data is stored in a hirarchical way so that you can search an element very fast. Every tree has at least a root-node which is the anchor of all subnodes. Every Node has pointers to 8 subnodes.

A C(++) structure of such a node will look like this:

```
struct octreenode
{
 void *data;
 struct octreenode childs[8];
};
```

where data point to any data and the childs point to other

octree nodes or point to NULL and indicate that there aren't any childnodes (such nodes are called leafes).

INSERTION OF DATA

Data is stored in a binary way. In our example we want to store RGB-tripples (colours). Now lets insert a colour into the tree. We start at the root-node of the tree and examine our RGB-Trippes. From each element we take the most significant bit combine them into a index-byte like this: 00000RGB where R,G, and B are the bits we have extracted. The higest value this byte could have is 7 (binary 00000111) and the lowest is zero (binary 00000000). This byte is an index of the child-nodes we have. Now we look at the child node pointed by our binary index. If it's NULL we create a new node and let the child-node point to it. We also initialize all fields to zero. Then we traverse down the tree and take a closer look to node childs. We've done the first step. Now we examine the next bits of our tripple and continue the traversal until we've used all 8 bits of our input-data.

COLOUR QUANTIZATION AND DITHERING

Reducing the number of colours or replacing specific colours is a very complex and difficult step in Image Magick, and that is the topic covered in this example pages. This includes reduction of the overall number of colours, dithering or patterning the reduced number of colours, generating bitmap images, and even handling Boolean (on/off) transparency.

This is so important that colour reduction or quantization often happens automatically just so IM can save to formats that can't hold all the colours available in an image, for example GIF, XPixmap, and XBitmap formats. Knowing this you can do the colour reduction yourself, so as to improve the overall resulting image stored in a particular file format.

Colour Reduction Introduction

Colour reduction is a very important aspect of ImageMagick. For example to convert a JPEG or PNG image containing millions of colours, into a GIF image containing a maximum of 256 colour, you really have to be able to reduce colours in a efficient and effective way. Often during an image format conversion, this happens automatically behind the scenes, but there are other times when you want to do this manually.

Reducing the number of colours in image is a typically a three step process:

- First you (may) need to survey the colours an image uses. Not only to see how many colours are actually used, but how often a particular colour is used. It is no good preserving one specific colour, if only a single pixel is using that colour, though sometimes you still need to do that.
- Next you need to somehow decide on the final colour set to which you want to limit your image. Sometimes you what the'best' set of colours for a specific image. Other times you may want something more general and global that can be used on any image.
- And finally you need to modify the image so as to only use the colours you have selected. Preferably you want it so the results will look good, or perhaps you want it so that it will compress, compare, or optimize well.

Basically while colour reduction is often automatically handled behind the scenes, you need to be aware that it is happening, and what its effects will be.

Colour Survey

This is probably the lest important, and while IM provides you with methods to perform a survey, it is rarely done by users for colour reduction. I will leave further discussion to the relevant section, Extracting Image Colours.

Colour Selection (Quantization)

There are four basic methods for the selection colours.

The IM Colour Quantization method ("Adaptive Spatial

Subdivision" using octrees) surveys the colours in an image. It then attempts to choose a specific set of colours to best match a specific image, within the limits given. With"-map" you can give IM your own set of predefined colours.

You can threshold all or specific colour channels of the image, essentially making each colour channel purely Boolean or on/off. That is each colour channel can be given a value of zero or MaxRGB (IM'Q' level dependant). This however only produces a minimal set of about about 8 colours. A very limited colour set. The final alternative is to mathematically divide up each colour channel into a set of colour levels or intensities producing a'uniform colour map'. That is a colour map with each channel set to a constant set of values or intensities. Threshold itself could actually be regarded as the minimal uniform colour map with only two values.

These four colour control methods: Quantization, Predefined Colour Map, Uniform Colours, and Threshold; all have their limitations, as you will see.

Here is an example of each of these four methods:

```
convert
colourwheel.png +dither -colours
32 colour_quantize.gif
convert
colourwheel.png  +dither  -map  colourtable.gif
colour_predefined.gif
convert
colourwheel.png +dither -posterize
3 colour_uniform.gif
convert
colourwheel.png  \  -separate  -threshold  50%  -
combine colour_threshold.gif
```

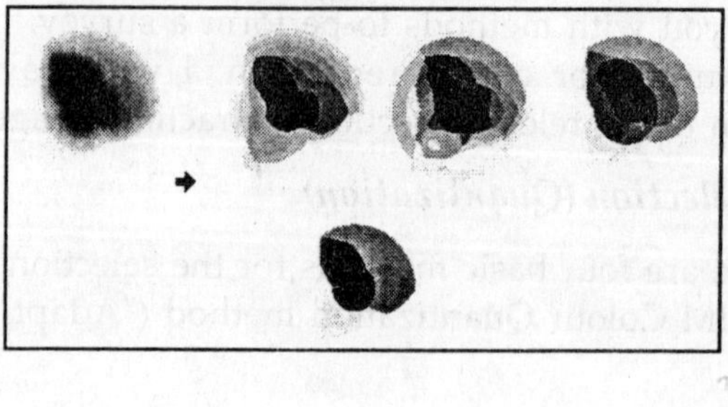

The number of colours in each of the final images is only a representative set, but approximately 32 colours in each case (except for threshold which only has 8). From this you can get an idea of what you can expect from each of them.

Only the first colour quantization, method actually will pick the colour to be used based on the the current image contents. As the test image is predominately white, a lot of lighter colours is selected.

All the other methods have a fix set of colours (according to the operators argument) regardless of the image that is being colour reduced.

The colour map"colourtable.gif" used in the above, is a set of 32 colours specifically picked for use in an old X Window Icon Library and is designed with cartoon like icons in mind.

Applying a Colour Set

Once you have a set of colours, the next problem is to apply the colour to an image so that it only uses the selected set of colours. This is known as'Dithering', and is named as such because of its,"should I pick this, or should I pick that?", nature.

Basically the idea of dithering is to place pixels of different colours near each other in such a way as to fool the eye into seeing more colours in the image than is actually used. That is the colour in that area of the image more closely matches the original colour of the image, because of the way the human eye'merges' neighbouring colours together.

This presents an excellent set of examples of the benefits of using a dithered pattern of pixels, when you have a limited set of colours.

Styles of dithering include:

- Direct Colour Mapping
- Random Dither
- Error Correction Dithers
- Ordered Diffused Pixel Dither
- Digital Halftone Dither

Direct mapping if the nearest colour in a given set is what

was shown above. Basically you get distinct areas of solid, unchanging colours. When this is applied to a image of slowing varying colour, such as a real life photo of sky, you get bands of colours. Not very good.

The only time direct colour mapping is usually thought of as acceptable is for logos, symbols, icons, and cartoon-like images. It is actually rarely an option. This is why you generally have to turn it off, if you do not want dithering in your images.

Dither however has its own problem. Once a image is dithered, a pattern of colours becomes part of the image. Once such a pattern is present, it is extremely difficult to remove. Also it is generally a bad idea to re-apply dithering to an image multiple times, as that just degrades the image.

Because of this, the examples will generally show you how to create un-dithered versions for each technique. So if you don't want dithering, you don't need it. It is also done so you can see what colour selections are being made before dithering hides that information.

Random dithering is the simplest dithering method created. It is also regarded as the worst possible dithering method.

However it has some special uses. Within IM only works with the colours threshold, so it is usually restricted to special case bitmap dithering. Error Correction Dithering is generally regarded as being the best method of dithering colours across images as it will produce the closest approximation to the original colour of areas in the image. It is also the only method that can dither anyset of colours, and as such can be used for all four colour reduction techniques.

However Error Correction Dithering has some serious problems, especially with regards to animations of images. The last two dithering techniques Ordered Diffused Pixel and Digital Halftoning is also regarded as a good method, and one that works well for animations, but it can not just use any set of colours, only a fixed set of uniform colourmap.

It also, however, provides a means of using a set of graded patterns in images, allowing you to produce interesting effects otherwise not easy to produce. All these aspects of colour reduction are important techniques, and with understanding you can improve the results of your image operations, beyond the generalised defaults that IM provides. It is well worth the studying.

Extracting Image Colours

Information about images, such as the number of colours used and the overall spread can be very important to programmes and scripts that are trying to make decisions about the best techniques to use. Here I look at some of the methods you can use to determine this type of information, and not just for colour reduction.

You can of course use a verbose"identify" to extract either the colour table the image is used (for index colour image formats like GIF), or the histogram of the colours used.

```
identify -verbose image.png
convert
image.png miff:- | identify -verbose    -
convert
image.png -verbose -identify null:
convert image.png -verbose info:
```

The output from any of the above verbose identification will not return the colour tables or histogram if there are more than 1024 colours! As such for large colourful images this is a hit or miss affair, and not recommended, though with an initial colour reduction it can still be useful.

The better way however is to generate a"histogram:" of the image and extract the comment that is included in the result.

```
convert
tree.gif -format%c -depth 8 histogram:info:
 16: (0,128, 0) #008000 green
422: (0,255,255) #00FFFF cyan
126: (46,139, 87) #2E8B57 SeaGreen
381: (50,205, 50) #32CD32 Limegreen
 59: (160, 82, 45) #A0522D sienna
 20: (255,255, 0) #FFFF00 yellow
```

The"info:" output fo rmat was added to IM v6.2.4. For IM versions before this use..

```
convert
tree.gif histogram:-| identify-depth 8-format%c-
```

The problem with these methods is that you are given a plain text output of the colours, which you will need to parse for your own needs.

However as of IM v6.2.8-8, the"**-unique-colours**" operator will convert an image into a smaller one containing just one pixel per unique colour found in the original image, all in a single row.

This means you can convert a image into a simpler colour table image, listing each colour present. The width of the image returns the number of colours, and if you need to actually list the colours you can output it to a"**txt:**" image format.

For example here is the colour table for the tree image:

```
convert
tree.gif-unique-colours-scale 1000%
tree_colours.gif
convert
tree.gif -unique-colours -depth 16 txt:-
  # ImageMagiC pixel enumeration: 6,1,65535,rgb
  0,0:  (41120,21074,11565)  #A0A052522D2D
sienna
  1,0:  (    0,32896,    0)  #000080800000 green
  2,0:  (12850,52685,12850)  #3232CDCD3232
LimeGreen
  3,0:  (11822,35723,22359)  #2E2E8B8B5757
SeaGreen
  4,0:  (65535,65535,22359)  #FFFFFFFF0000
yellow
  5,0:  (    0,65535,65535)#0000FFFFFFFF
cyan
```

The order of the colours in the single row produced by the"-unique-colours" operator, is undefined.

This reduced colour table is also very important as a way of storing a colourmap of generated colours in a very small file. Such maps are particularly important for the"**-map**" colour reduction operator.

The average colour of an image can be found very quickly by using"**-scale**" to reduce an image to a single pixel. Here for example is the average colour of the built-in"**rose:**" image. I output the colour using the FX Escape Format which returns a colour string that can be used directly IM without change.

```
convert rose: -scale 1x1\! -format'%[pixel:u]'
info:-
rgb(146,89,80)
```

The problem with using the"**%[pixel:...]**" FX Escape is that it may return a colour name such as'**white**' or'**silver**' instead of a RGB value. However you can simulate this by using three FX Escapes to return the actual RGB values at the bit depth wanted.

For example:

```
convert
rose:-crop1x1+12+26\-format'%
[fx:int(255*r)]
[fx:int(255*g)]
[fx:int(255*b)]
info:-
136,129,210
```

As you can see you can simplify the expressions by using"**-crop**" to cut out a single pixel that you may be interested in. Here for example is the same pixel using TXT: pixel enumeration. This may be better for scripting purposes.

```
convert
rose: -crop 1x1+12+26 txt:-
# ImageMagick pixel enumeration:1,1,255,rgb
0,0: (136,129,210) #8881D2 rgb(136,129,210)
```

As of IM v6.3.9 there are a number of new"**-format**" escapes that can be useful to extract more specific information about images without needing to parse a verbose"**identify**" or"**info:**" output. For example you can get the average red channel colour by getting the'**%[mean]**' greyscale value from of the images, red channel image.

```
convert
```

```
rose: -channel R -separate -format'%[mean]' info:
37448
```

Colour Quantization Operator

The primary work horse of colour quantization, and what is used internally for all automatic colour reduction, is the"-**colours**" operator.

This implements a"Adaptive Spatial Subdivision" colour reduction algorithm, and is an extremely good colour reduction algorithm.

Here is a typical example, I have a image of a'colourwheel' image containing a lot of colours, and we ask IM to reduce the number of colours down to only 64 colours, using various dither methods.

```
convert
colourwheel.png -dither None -colours 64
colours_64_no.gif
convert
colourwheel.png -dither Riemersma -colours 64
colours_64_rm.gif
convert
colourwheel.png -dither FloydSteinberg \
-colours 64 colours_64_fs.gif
```

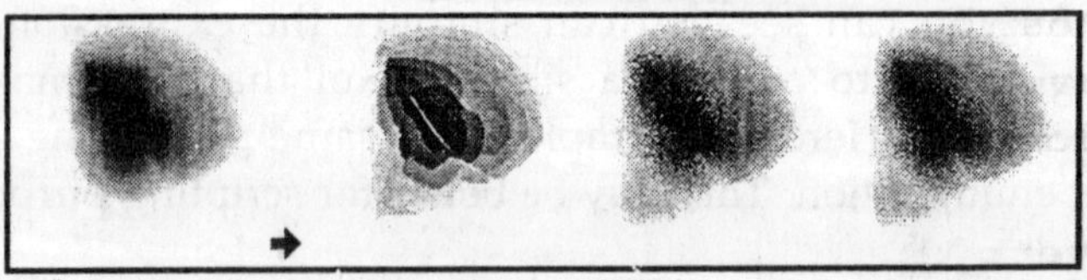

You can also see IM will by default use a'dither' to shade the colours over the image. This prevents the sudden changes in colour over smoothly changing gradients. If you turn of dithering (using'**None**' or a"**+dither**" setting) you can clearly see what colours were merged together to generate what IM regarded as the best set of colours for this specific image. You can also see the sudden colour changes that gradients of colour will produce is dithering was not done.

Of course this image uses a lot more colours than what most images use. As such while a 64 colour limit is often acceptable for many images, it is completely unacceptable for this image. In other words colour quantization tries to find the best set of colours for a particular image.

Colour Quantization Internals

The process of selecting the limited number of colours to use in an image is called Colour Quantization, and is a very complex process involving a number of factors. A full technical description of it is given on the ImageMagick web site Colour Reduction Algorithm. However I'll try to example some of the more important aspects of this here.

Probably the biggest factor is the actual colours used in an image. It is no good picking a particular colour for an image if their are very few pixels that are'close' to that colour. As such the colour choice depends not only on the colours used in an image, but the number of pixels'close' to the colour.

```
convert
-size 4x1 xc:blue -draw'fill red point 0,0' \
-scale 20 colours_rb.gif
convert
-size 4x1 xc:red -draw'fill blue point 3,0' \
-scale 20 colours_br.gif
convert
colours_rb.gif -colours 1 colours_rb2.gif
convert
colours_br.gif -colours 1 colours_br2.gif
```

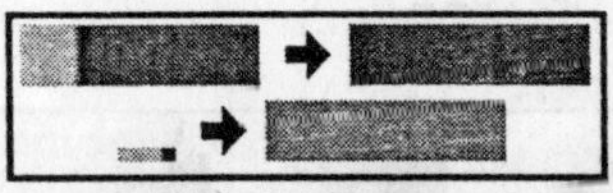

As you can see the single final colour depends not only on the colours present, but the amount of each colour in the image. Another thing that should be noted is that colour quantization is not uniform over a linear gradient. This is to compensate for the typical'gamma' correction that is present in images.

```
convert
-size 20x640 gradient: -rotate 90 gradient.png
convert
```

```
gradient.png +dither -colours 5
colours_gradient.gif
```

```
FUTURE: Just what are the effects of the"-
treedepth" setting?
Mail me if you know
```

Colour Quantization and Colour Space

The other big influence on what colours are selected is defining exactly what we mean by colours that are'close' or'nearby'. This is defined by the colourspace used for the quantization (colour selection), and is (as of IM v6.2.8-6) controlled by the"**-quantize**" colourspace setting.

The"**-quantize**" setting becomes particularly important when a very small number of colours are chosen. To demonstrate, lets reduce a standard'colourwheel' image using various different colour spaces and defining different'colour distances'.

```
for S in RGB CMY CMYK GRAY YIQ YUV OHTA XYZ HSL
HSB HWB; do \
convert
colourwheel.png -quantize $S +dither -colours
16 \
-fill black -gravity SouthWest -annotate +2+2
$S \
colours_space_$S.gif; \
done
```

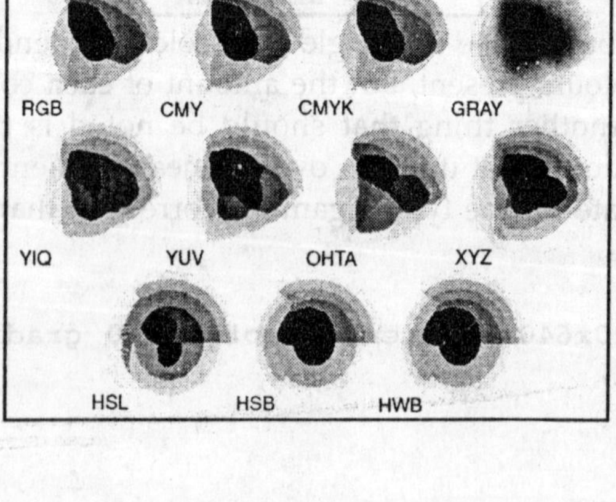

As you can see the colours chosen depend heavily on how the colourspace is organized. The RGB (Red, Green, Blue) colour cube will generally result in at least the colours close to the primary colour being picked.

Cristy notes that the CMYK colourspace acts like RGB colourspace because of the way a extra colour channel is handled internally by ImageMagick. Both an images colourmap and the CMYK black channel occupy the same data pointer. If we permit CMYK quantization it would overwrite the colourmap indexes, and thus the black channel, which is not good when we later try to dither/replace the colours.

The CMY quantization colourspace should not have the four channel colour problem of CMYK, but for some unknown reason it is still producing an RGB quantization pattern, rather than a CMY pattern. Attempts to fix this have failed, as the cause of the problem is known at this time.

The RGB colour space is particularly good at picking colours for cartoon like images and icons, but is actually a bad colour space for general picture-like photos.

On the other hand YIQ, YUV are designed to produce more natural'pastel' and'mid-tone' shades of colours that are much better suited for photographs and images of the real world involving subtle shades of colours.

Cristy, the ImageMagick Coordinator, notes that the most commonly used colourspace should be YIQ as it gives the best result when reducing a photograph of say a persons face to a limited colour image such as GIF. I also agree with that recommendation.

The special colour spaces involving a'Hue' channel, such as HSL, HSL (Hue, Saturation, Brightness), and HWB (Hue, White, Black), all have a cyclic colour wheel representation of colour as part of its colour space. Actually it was using a HSL colour space what was used to generate this colour wheel. Generating a Colour wheel.

At the time of writing, the distance algorithm IM uses does not take into account the cyclic nature of the'Hue' or colour of the colourspace, so a strong discontinuity occurs along the'red' path. The result is that currently these colour spaces do not generate many red colours in the colour quantization process. This has been reported.

Finally there are a larger number of some very weird colour spaces that are used for more specialized purposes. I have shown three of these above.

In older versions of IM (specifically IM version 5) the colour space that was used for quantization was set with the"-**colourspace**" option. However in IM version 6 this operator is used for modifying how images are stored in memory, and as such is not a setting for colour quantization.

As such in IM v6.2.8-6, the"**-quantize**" setting was provided to do this job. However it is only as setting for the"-**colours**", Colour Quantization process. It will not do anything for the replacement and dithering of colours using operators such as"**-map**" and"**-posterize**", or the various dithering techniques.

For a complete list of the colour spaces available see the"-**colourspace**" operator. You can see more effects of the colourspace on colour selection by looking at the examples on Random Spots of Solid Colour. There colour quantization is used to reduce the number of colours in a randomized image using various colourspaces.

Quantization does NOT Preserve Colours

We see that in all the above images a pure-black colour is never actually picked by Colour Quantization. The only black that appears was added later as part of the labeling of the image.

Even the'**GRAY**' colour space image did not produce a pure-black colour. In fact none of the images contains any of the primary or secondary colours, such as: red, blue, green, cyan, magenta, or even white! I also doubt a specific'named' colour was picked either.

First, a'**black**' colour was generally not selected in the above examples, usually because as their is very little black in the original image, so the Colour Quantization generally did not worry too much about dark colours. In fact it generated more of the lighter colours as these are more common in the image.

Secondly, as quantization is trying to pick colours that are close to the maximum number of existing colour pixels in an image, this is best achieved by NOT matching a'pure' primary or secondary colour as these are the always at the extremes of the colour space being used. An'off' colour will tend to match more colours than a'primary' colour, so these more often selected.

As such even though the original image used contained a lot of'**white**' pixels, after colour reduction a near-white colour was chosen to better represent all the whitish colours. As such not even white was chosen as on of the colours in any of the above examples.

As of IM version 6.3 the Colour Quantization function was modified to try to include colours that are common in the original image. As such if an image contains a area of a single colour, that colour will be included in the final colour map. This improves the situation somewhat, especially for'cartoon' like images or images on a solid colour background.

Specific Colour in Colourmap Solutions

At the moment there is only a few ways of fixing this'specific colour' problem.

One way is to quantize the image as normal, but then and output the generated colour map (using"**-unique-colours**". Now you can adjust that colour map so your white is really white. You can now Map the new colour map to your image.

The colourmap may no longer be the BEST colours for the image, and some other colours probably should also be adjusted, but it will be close to the colourmap that you wanted.

Alternatively, append large patches of the specific colours wanted to the image, before using"**-colours**". Those colours will hopefully become included in the final colour map (and all other colours adjusted to fit better).

If this works, you can then just Crop the image to remove the extra colour patches added. If it doesn't, then IM should have at least added a colour close to the wanted'specific colour' so only very slight adjustments of just those colours should be needed before you again Map the colours to the original image yourself.

If you try this, with success or failure, please mail me. Ideally, what it would like to see added to IM is a way to specify a small number of specific colours, that must be part of the final colour map, and then somehow ask IM to pick the best colours for rest of the colours in the colour map, for a specific image.

Colour Quantization and Transparency

ImageMagick by default not only generates fully opaque colours, but also attempts to generate semi-transparent colours. In this way, images containing transparent shadows or other overlay effects will not loose those effects.

However as IM v6.2.6, colour quantization that involves transparency was modified so as to treat all fully-transparent colours as being the same colour. This is a linear modification, so colours which are only half-transparent are also thought to be closer together than if they were fully opaque.

Because of this modification IM Colour Quantization will still generate semi-transparent colours, but will concentrate more on the opaque colours and less on the fully transparent colours in the image.

Colour Quantization and Colour Space

```
convert xc:red xc:yellow xc:green1 xc:cyan xc:blue
\
+append -filter Cubic -resize 100x100\! -size
100x100 \
gradient: +matte -compose CopyOpacity -composite
alpha_gradient.png
convert alpha_gradient.png +dither -colours 256
alpha_colours_256.png
convert alpha_gradient.png +dither -colours 64
```

```
alpha_colours_64.png
convert alpha_gradient.png +dither -colours 15
alpha_colours_15.png
```

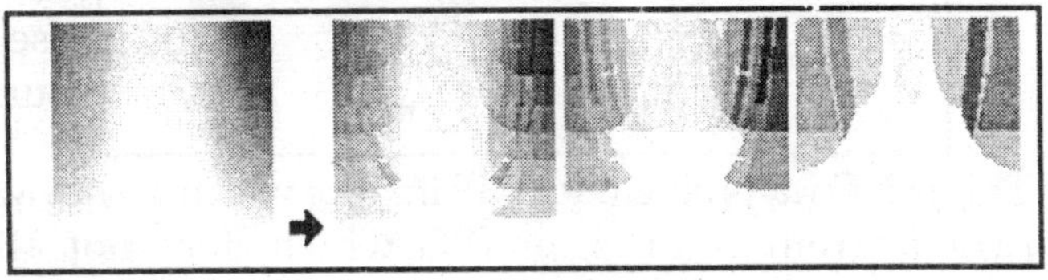

As you can see, when we ask IM to reduce the number of colours needed by this image, it created a lot more opaque colours and used fewer highly transparent colours for the more translucent parts. The result is a very good spread of colours selected, especially when the number of colours is very small.

However just as I pointed out above, not only do primary colours not get picked, but the fully-transparent colour will also not get picked for exactly the same reasons. In actual fact even fully-opaque colours will not get picked! In other words every colour in the colour quantized images in the previous example is semi-transparent.

Of course as of IM v6.3, and the'common colour' bug fix (see Quantization does NOT Preserve Colours above), that is less likely to happen if the image contains a lot of opaque and fully-transparent colours, which is commonly the case.

As some images can contain a lot of semi-transparent colours, such as images involving smoke or shadows effects, you may like to do a trial run, to make sure a fully-transparent colour is selected for inclusion in the resulting image. You can then map the most-transparent colour to fully-transparent, and do the Colour Mapping yourself.

If you really want to be sure you get both fully-opaque and fully-transparent colours in the resulting image you can Normalize OR Contrast-Stretch the alpha channel. For example here it is ensure the main colour selects are made opaque by using"**-contrast-stretch**". Though this is probably a little heavy handed for a more normal situation.

```
convert alpha_gradient.png +dither -colours 15
\
```

```
-channel A -contrast-stretch 10%
alpha_colours_15n.png
```

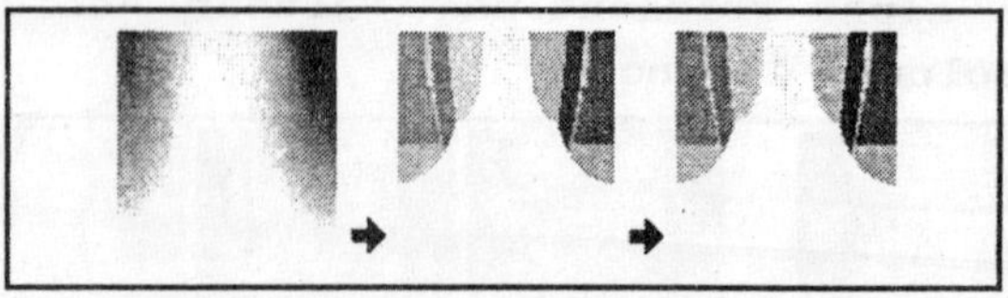

This is NOT a problem for GIF images which do not allow semi-transparent colours, or JPG which does not allow transparency, or even PNG which do not need quantization to save properly. It only becomes a problem in special cases where you may force a colour reduction in an image when lots of semi-transparent colours are involved.

Remember for the GIF format saving semi-transparent colours is a useless endeavor. As such if you plan to do colour quantization yourself for such an image format, you need to tell IM to ignore image transparency when generating its reduced colour set. you can do that by using the special"-**quantize**" colour space setting of'**transparent**'.

```
convert alpha_gradient.png -quantize transparent
\
+dither -colours 15 alpha_colours_15qt.png
```

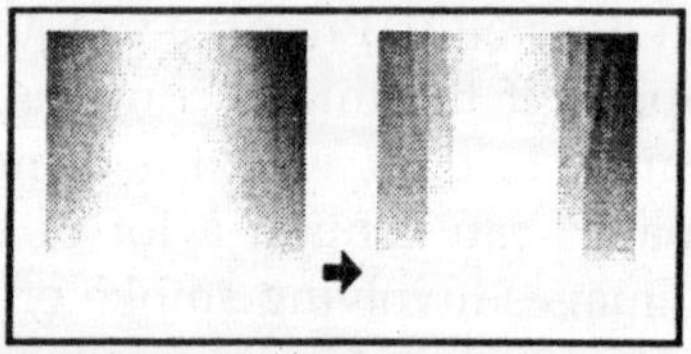

Notice how the colour quantization completely ignored the transparency of the colours, and did not touch the images alpha channel at all. This means you can process the alpha channel in a more appropriate way for your image, completely separately to the other colours. In fact you can do so either before or after using"**-colours**" without problems. It will make no difference in the result.

This quantization colour space is thus recommended when reducing the number of colours for an image you plan to save to a format with Boolean or no transparency, such as GIF or XPM image formats.

If you count up the number of colours generated you will also see that it generated exactly the number of colours requested. As such if you you also need a fully transparent colour (likely) then you need to reduce the argument of"-**colours**" by at least one, to leave space for it in the images final colour table.

Thus to handle the GIF file format 256 colour colour table limit, you will need to reduce colours to 255, and not 256, leaving the extra space for the fully-transparent colour index, as defined by the"**-transparent-colour**" setting. Adjust this for a smaller colour table sizes.

This quantization behaviour is automatic when IM saves to the GIF file format, but is important when you need to DIY the quantization yourself while generating global or shared colour tables. Of course you do still need to handle the semi-transparent pixels, so they are correct for what you want your image to look like.

Here are some examples of dithering just the alpha channel to just a Boolean or on/off setting, without effecting the rest of the colour channels in the image.

```
convert alpha_gradient.png \
-channel A -threshold 50% alpha_dither_ threshold
  .gif
convert alpha_gradient.png \
-channel A -ordered-dither checks alpha_dither
  _checks.gif
convert alpha_gradient.png \
-channel A -ordered-dither o8x8 alpha_dither_
  ordered.gif
convert alpha_gradient.png \
-channel A -ordered-dither h8x8a alpha_dither_
  halftone.gif
convert alpha_gradient.png -channel RGBA -separate
  \
\(+clone -monochrome \) \
-delete 3 -combine alpha_dither_monochrome.gif
convert alpha_gradient.png -channel RGBA -separate
\
```

```
\(+clone -dither FloydSteinberg -monochrome \)
\
-delete 3 -combine alpha_dither_monochrome_fs.gif
convert alpha_gradient.png -channel RGBA -separate
\
\(+clone -map pattern:gray50 \) \
-delete 3 -combine alpha_dither_map.gif
convert alpha_gradient.png -channel RGBA -separate
\
\(+clone -dither FloydSteinberg -map
pattern:gray50 \) \
-delete 3 -combine alpha_dither_map_fs.gif
```

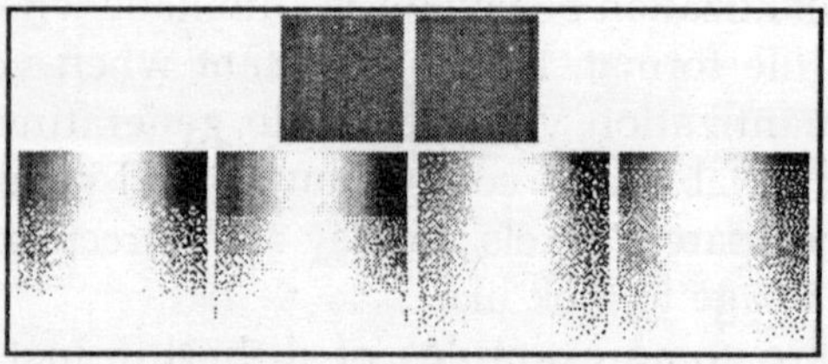

When dithering a copy of the Alpha Channel, so you can dither it using either"**-monochrome**", or"**-map**", make sure the image is a pure grayscale image, and not a shape mask containing transparency. If you don't you will probably end up with non-linear effects from the alpha channel still being present.

There are a number of ways of extracting and restoring the alpha channel from an image, as a gray-scale mask so you can dither it. The above uses the Channel Separation and Combine to do this. Other methods use Alpha Extraction with CopyOpacity Composition.

ERROR CORRECTION DITHERING

As discussed in the introduction an error correction dither is generally regarded the best choice for producing the truest representation of the original image with a reduced colour set. It also limits itself to any pre-defined palette of colours, whether it was user supplied, or as determined by the IM colour quantization routines.

Because of this it is the logical default choice for general colour reduction as provided by the IM operators,"**-colours**","**-map**","**-posterize**" and"**-monochrome**".

E-Dither Methods

As of version 6.4.2-9, IM now provides more than one type of dithering style or method, which can be selected using the"**-dither**" setting. Before this IM was limited to a variation of the Riemersma Dither, or Hilbert Curve Dither. which you can set using"**-dither Riemersma**".

Now you can also select a Floyd-Steiberg Dither using"**-dither FloydSteinberg**". You can see what types of dither methods has been implemented in your version of IM using.

```
convert -list dither
None
FloydSteinberg
Riemersma
```

For example here is the colour wheel dithered using different dithering methods:

```
convert colourwheel.png -dither Riemersma -colours 16 dither_riemersma.gif
convert colourwheel.png -dither FloydSteinberg -colours 16 dither_floyd.gif
```

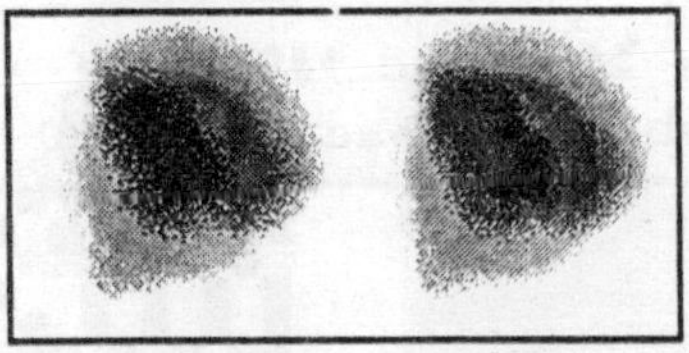

As you can see the Floyd-Steinberg dither produces a much more uniform dither pattern than the default Riemersma dither. The biggest difference between them is how each of them distributes the'colour error' between neighbouring pixels. So lets have a look at just how a E-Dither works.

WORKING OF E-DITHER

Write In Progress

The specific method used by IM for general dithering is

a variation of the"Hilbert Curve Error Correction Dither". This is actually a very good dithering technique, very well defined and reasonably quick. For a full description (and a very similar variation) Riemersma Dither.

Basically each pixel in the image is looked at in a very complex path known as a'Hilbert Curve'. The pixel is assigned the colour closest to that pixels value, and any difference between the pixels original colour and the selected colour, is saved and added to the next pixels colour values (which is always a eighbouring pixel) before a new colour is again selected. In this way any colour variations between selected colours and the images original colour is distributed to the other pixels in the same area. The result is that while only specific colours will be assigned to the final image, the same basic overall colour for that area will closely match the original image.

For example, here is a small grey image that I asked IM to dither using a set of colours that does not include the original colour. The resulting image is magnified so you can see the individual coloured pixel assigned.

```
convert -size 10x10 xc:'#999999' -scale 80x80
dither_not.gif
convert -size 10x10 xc:'#999999' \
-map colourtable.gif -scale 80x80 dither.gif
```

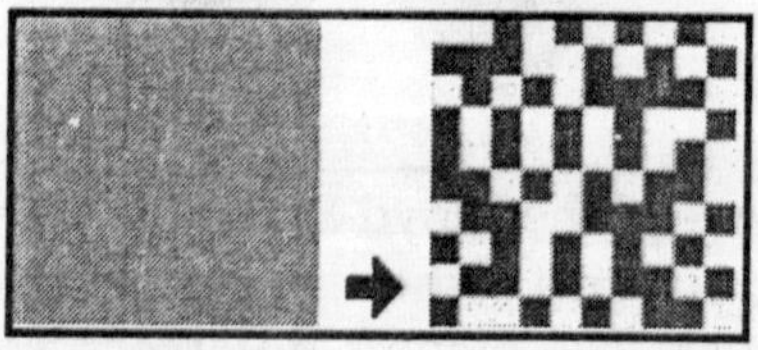

As you can see the as the original images colour was not in the specified colour map, the original colour is approximated using a pattern of the three nearest colours that were in the given colour table.

If we were to average the colour generated by the above dither pattern, we would get the colour #999A99, which is very close to the images original average colour of #999999 and that is the whole point of the dither pattern that was produced.

However as the'path' used to assign colours is complex (though generally remains in the local area), the colour assignments produce an essentially random pattern. It isn't technically random however as the same image will produce the same pattern, but the results may as well be random, or at least pseudo-random.

The"F-S" dither is actually only one (the first one) of several'Rasterized E-Dithers' that has been developed since its inception in the early 1970's. It is also probably the most widely implemented, even though it is not regarded as the best one. As of IM v6.4.3 it is also directly available in IM, and is implemented so as to follow a'serpentine' path row-by-row from the top of the image to the bottom.

```
convert -size 10x10 xc:'#999999' -dither FloydSteinberg \
-map colourtable.gif -scale 80x80 dither_fs.gif
```

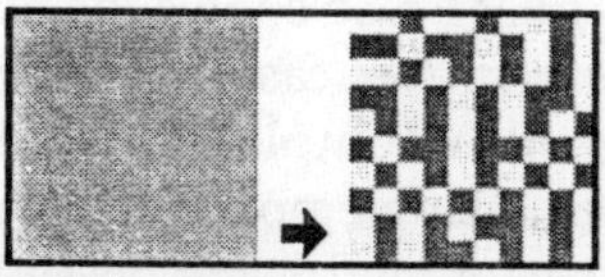

The"Floyd-Stienberg Dither" in particular I find produces a more'hash' like pattern of pixels than the"Hilbert Curve Dither", and was in fact designed to do so.

Such a regular pattern can make a low level manual clean up of small colour icon images, a lot easier. This was something I did a lot of in my past for Anthony's Icon Library, but that type of thing is not often needed anymore, except posibly for small monochrome images.

E-Dither Problem - Change Sensitive

One of the biggest problems you face when you use an error correction dither is that you get an essentially random pattern of pixels, that is also highly sensitive to changes.

Here for example take the original grey image and replace one pixel with a different colour before dithering it again. The result is a complete shift of the dithering pattern in every pixel that is further along the path followed asr.

```
convert -size 10x10 xc:'#999999' -draw'fill #C28
```

```
point 2,2' \
-map colourtable.gif -scale 80x80
dither_modified.gif
compare dither.gif dither_modified.gif
dither_difference.gif .
```

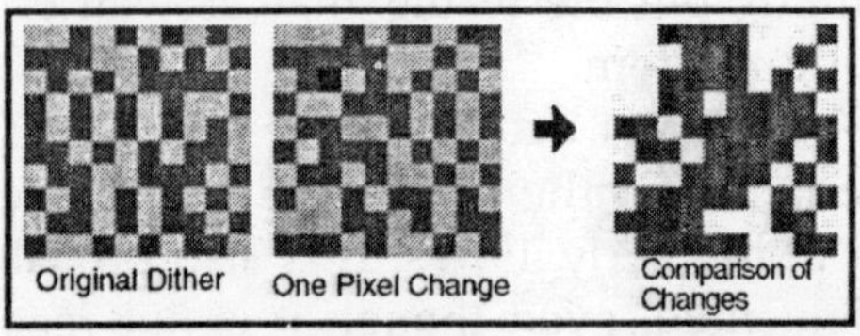

As you can see just adding a single pixel to the image resulted in the dither pattern changing dramatically! It only takes a single bit change for the resulting image to become different, even though overall look of an image (when not enlarged) is still basically the same (which is the purpose of a good dither algorithm after all).

The"**compare**" image also shows the extent of the change in the dither pattern. In this case, approximately 80% of the pixels were assigned a completely different colour. In a Hilbert Curve Dither, a single pixel change actually will result in every pixel that comes later being possibly different, which means from 0 to 100% per cent of the dither pattern could be different. It just depends on where in the complex Hilbert curve the change occurred. The Floyd-Steinberg dither however only progresses though the image in one direction, and as such a single pixel change will modify the pattern only to one side of the change.

```
convert -size 10x10 xc:'#999999' -draw'fill #C28
point 2,2' \
 -dither FloydSteinberg -map colourtable.gif \
 -scale 80x80 dither_fs_modified.gif
 compare dither_fs.gif dither_fs_modified.gif
dither_fs_difference.gif
```

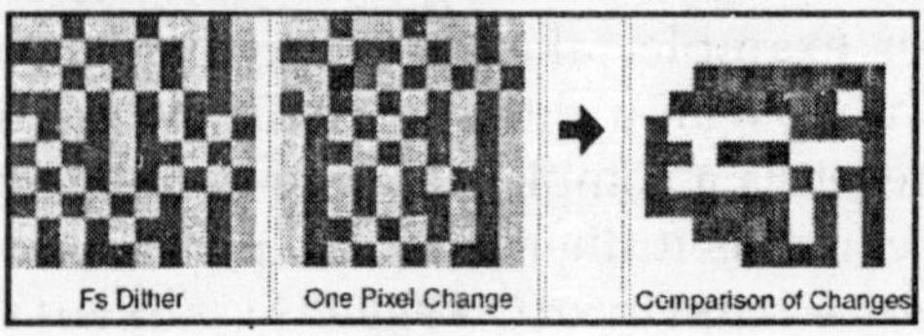

As you can see it has exactly the same problem. A single pixel change causing an almost complete change in the dither pattern for the areas of the image processed after that pixel. That is from that row downward.

For a single image the resulting pattern of the dithered colours is unimportant. The average colour of the pattern should give the image the appropriate colour for that area of the image. But when you have an animation in which one image is followed by other very simular images, with large areas of constant colour, the changing dither pattern becomes highly noticeable and irritating as a low level background'noise'.

```
 convert -size 80x80 xc:'#999999' \
 \(+clone -draw'fill #C28 point 2,2' \) \
 \(+clone -draw'fill #28C point 2,2' \) \
  -map colourtable.gif -set delay 50 -loop 0
dither_anim.gif
  convert dither_anim.gif -crop 10x10+40+40
+repage \
 -scale 80x80 dither_anim_magnify.gif
```

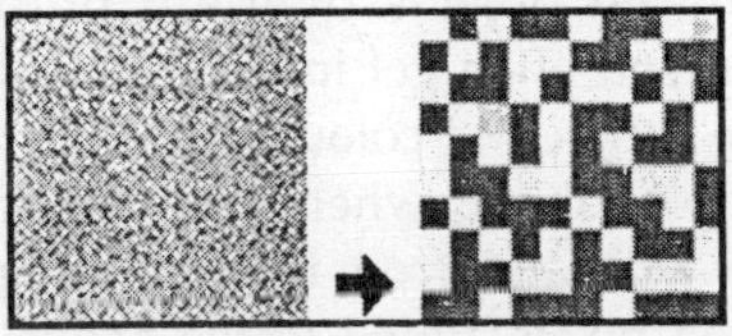

As you can see you get a sort of churning background to the image, caused by the pseudo-randomness generate by the E-Dither. In most cases the colours used are close enough together so as not to make this'dither noise' visible. But when the dithering colours are visibly different (in this case forced by the use of a colour map) it definitely becomes an issue.

The change in pattern also causes problems in optimizing animations. That is a different pattern means that simple frame optimization fails to reduce the size of frame overlays. For one solution see fuzzy colour optimization, though that only works when the churn is using very similar colours.

Unlike other dithering methods (such as threshold, and ordered-dither) the"-channel" setting does not effect colour quantization, or error correction dithers. Basically has it has no place in how these image operation work.

Ordered dithers do not have any of these problems, containing changes to the immediate local area of the change. Unfortunately are also generally limited to using a mathematically derived colour set. (Ordered Dither using a Uniform Colour Map). I am working on an simple ordered dither (very simple) that can use a user provided colour map.

E-Dither Pixel Speckling

Another problem with E-Dithers is that they can produce the occasional odd-coloured pixels in areas which would otherwise be fairly uniform in colour. For example the occasional green, pixel in a grayscale image. Or as in the examples below, a white pixel in a areas of otherwise plain flat blue colour.

This is especially the case in large images which containing objects with large numbers of colours, and other areas of plain solid unchanging colours. This is especially typical of coloured objects overlaid onto flat coloured backgrounds, as you often get in diagrams and drawings.

You can see such a odd-coloured pixel in the enlargement of the test examples above, where an extra light purple pixel was added quite a distance from the small single pixel change. The odd-coloured pixels added to the above is not however readily visible and the colour map does cover the image rather well, so the odd pixels are reasonably close to the normal three colours used for dithering the image.

The best solution is to switch to some other image format that does not have a limited colour table. For example convert you GIF format image to PNG. This will avoid the need for colour quantization (reduction) and hence the need to dither the reduced colours.

The next solution is to replace the use of E-dither with some other dithering method, that'localizes' any errors, such asOrdered Dithering. However that is currently not easy thing to apply in IM at this time. Better Ordered Dither Results, for one such method until a more general one is found.

If switching to another image format, or using a different dithering method is not practical (and often isn't), then you are left with attempting to fix the situation for that specific image.

The best fix, to this is to somehow insure you have other colours just outside the large group of colours that is causing the E-Dither error accumulation. However normal Colour Quantization does not do this. It tends to pick a set of average colours representing colour groups. What is needed is extra colours that picked fence the edges of the a large colour group, rather than an average colour.

Here for example, I used a circle rather than a square, so that not only is a pure white colour added, but a number of white-blue colours also. These were added automatically due to the anti-aliasing of the circle edges, to smooth its look.

```
convert -size 100x60 xc:SkyBlue \
-fill DodgerBlue -draw'circle 50,70 15,35' \
 -fill RoyalBlue -draw'circle 50,70 30,45' -
blur 0x5 \
-fill white -draw'circle 50,45 40,40' \
-colours 64 speckle_fixed.gif
```

And a magnification of the same area as before:

```
 convert speckle_fixed.gif -crop 15x15+85+0
+repage \
-scale 90x90 speckle_fix_mag.gif
```

As you can see the additional colours provide the extra colours, just outside the blue-cyan gradient. Now while these extra colours will mean there are fewer colours available for the actual gradient, they do provide other colours in a blur-white gradient that will allow the E-dither to correct itself sooner, and more often, before the accumulated error can grow too large.

That is not to say we have not preventing E-dither speckling, just provided better colours for the dither algorithm to work with. If you study the magnified portion of the image you will still see a speckle pattern, but the colours are closer to the background colour, and there are a lot more of them producing a more even, localized spread of speckles.

Another way is to generate our own colour table, perhaps based on the one IM generated, and add the appropriate colours to prevent the error accumulation. This is however not an easy thing to do. For this specific image, one way to prevent'speckling' is to generate and dither the background, with slightly less colours that is needed, then overlay the white box.

```
convert -size 100x60 xc:SkyBlue \
-fill DodgerBlue -draw'circle 50,70 15,35' \
-fill RoyalBlue -draw'circle 50,70 30,45' -
blur 0x5 \
-colours 63 \
-fill white -draw'rectangle 40,40 60,55'
speckle_perfect.gif
```

This will add a'white' colour to the image, but the background will not have any speckling effect, as white was not available when it was dithered using the error correction dither. The result is an image with exactly 64 colours, and no speckling at all. However this is very dependant on the image and what you are trying to achieve, so is not a general solution to the specking problem.

A more general alternative to adding extra colours is try to remove the speckles from the final dithered image. That is clean up the image in some way. However this is itself a tricky problem, as you do not what to remove pixels which are part of the normal dithering pattern. What we need is to find colour pixels which are somehow very different to all the the colours surrounding it, but which are also well isolated from all other similar colours, by some distance.

Monochrome Dithered Bitmap Images

The"-monochrome" operator is a specialized form of both the"-colours" operator to generate a bitmap image. It is as such an ideal operator to demonstrate not only'Hilbert Curve Dithering', but also have a closer look at colour selection.

The operator dithered the image solely based on their grey-scale brightness'intensity' or'level', however it doesn't dither the whole grey-scale range directly but thresholds the most extreme values to their maximum values.

We can see this by asking IM to dither a gradient image.

```
convert -size 15x640 gradient: -rotate 90 \
-monochrome monochrome_gradient.gif
```

As you can see the gradient only has about the middle 50% of its colours dithered by the"**-monochrome**" operator. Special thanks goes to Ivanova <flamingivanova@punkass .com> for pointing out this interesting fact of the was IM works. If you like to dither using the whole grey-scale range, you can use the"**-map**" operator using a pure black and white colourmap (supplied by a built-in pattern image).

```
convert logo.png -map pattern:gray50
mono_map.gif
convert -size 15x640 gradient: -rotate 90 \
-map pattern:gray50 mono_map_gradient.gif
```

By more careful selection of colours using"-map" you can effectively produce the same'threshold' range as used by the"-monochrome" operator, or any other threshold range you like.

```
convert xc:gray20 xc:white +append
ctrl_colours.gif
convert logo.png -colourspace Gray \
-map ctrl_colours.gif -normalize
mono_map_ctrl.gif
convert -size 15x640 gradient: -rotate 90 \
-map ctrl_colours.gif -normalize
mono_map_grad_ctrl.gif
```

What"**-monochrome**" actually does is to first convert the given image first into a grey scale image, after that it performs a two colour'Colour Quantization', to decide the threshold colours to dither the image with. This is what the next section of examples will explore.

Two Colour Quantization

Rather than picking the two control colours yourself, you can use colour quantization to pick the best two colours in the image by using the"-colours" operator.

```
convert logo.png -colours 2 -colourspace gray
-normalize \
 colours_monochrome.gif
```

However the result will not be the same as using"-**monochrome"**, as we didn't convert the image to grey-scale first.

Instead the image was dithered directly between the two non-grey colour values chosen. That is the best two colours is selected to dither the image with, rather than two gray-scale brightness levels. Consequently it will produce a better result for say image that only uses colours of about the same gray-scale'level'. Here for example we use"**-colours**", as well as a"**-monochrome**" bitmap dithering operator, on a red-blue gradient. As you can so you can see that the results are not the same.

```
convert -size 20x640 gradient:red-blue -rotate
90 gradient_rb.png
 convert gradient_rb.png -colours 2 -colourspace
gray \
 -normalize colours_threshold.gif
 convert gradient_rb.png -monochrome
mono_threshold.gif
```

The"**-monochrome**" operator in the above failed to find any differences to bitmap dither as both blue and red are very nearly the same intensity. Using a"**-colours**" quantization method however had no problem in finding acceptable colours to dither between.

You can also see that only the middle portion of the colours were dithered. This is due to the colour quantization picking colours in the middle of two colour'clusters' it selected. Colours on the'outside' of the selected colours in is thus effectively threshold directly to that colour without dithering.

This demonstrates that colours on the outside of the quantization colour space does not get dithered, though this fact is difficult to make use of in a practical way. By setting the"-colourspace" to gray-scale before quantization, you will reproduce the internal operation of the"**-monochrome**" operator.

```
convert logo.png -colourspace gray -colours 2
-normalize \
monochrome_equivelent.gif
```

And finally, by turning off the dithering, you can produce a more automatic separation of the colours in the image than produced by using a fixed"-threshold" setting.

```
convert logo.png -colourspace gray +dither -
colours 2 -normalize \
threshold_two_grays.gif
```

If you remove the"-colourspace" for the colour quantization stage of the image processing, you can threshold an image based on the best colour separation (rather than greyscale colour separation) possible for that image.

```
convert logo.png +dither -colours 2 -colourspace
gray -normalize \
threshold_two_colour.gif
```

Pre-Defined Colour Maps

As shown above"**-colours**" attempts to choose an optimal limited set of colours with which to represent an image. With"-map" you provide IM with the final set of colours you want to use for the image, weather you plan to dither those colours, or just replace the ones with their nearest neighbours.

For example here is of the colours used in the IM logo to a predefined map of named X window colours. The default is the'Riemersma' dither, but as of IM v6.4.4"**-dither**" was expanded to allow the selection of other dither methods such as'**FloydSteinberg**'. You can of course still turn off dithering using the"**+dither**" option.

```
convert logo.png -dither None -map
colourtable.gif map_logo_no.gif
convert logo.png -dither Riemersma -map
colourtable.gif map_logo_rm.gif
convert logo.png -dither FloydSteinberg \
-map colourtable.gif map_logo_fs.gif
```

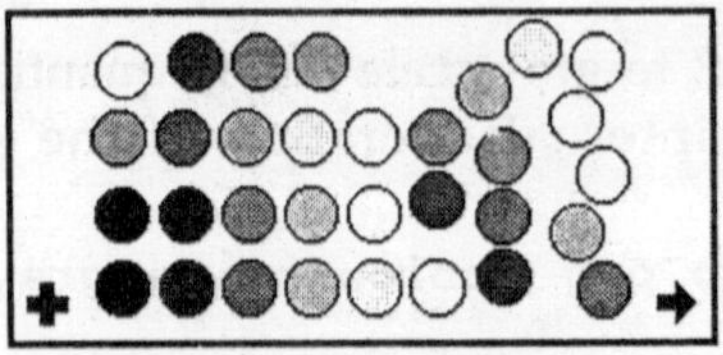

As you can see IM attempted to do a reasonable job of representing the image using just the given colours, though the results is nowhere near as good as the image you get if you allowed IM to select the colour set to use.

Mind you this"**colourtable.gif**" image was never designed for dithering images, but as a colour set for designing cartoon-like colour icons for older more primitive X window colour displays.

Mind you IM's selection of'**LightSeaGreen**' instead of say'**DodgerBlue**' for the wizards jacket is rather surprising and is probably due to the use of RGB colourspace for colour distancing. Unfortunately you do not have a choice about the colour space for actual colour internal map/dither operation.

Also note that the final image did not use all 32 colours provided by this map, though more of the colours in the map will be used when some form of dithering was enabled (27 colours and 24 colours respectivally), than when it was turned off (17 colours).

This last example shows just how important selecting a good colourmap is. Because of this I recommend you let IM optimize the colour selection used in an image using the"-**colours**" operator, and modify that to suit your needs, unless you have more pressing reasons not to do so.

One final point, while you can specify a colour space in which"**-colours**" will find the best set of colours, you currently can NOT define a colour space for the colour mapping or dithering phase. All my experiments seem to show that the colour set is applied (both error correction dither and nearest colour replacement) based on RGB space. The"**-quantize**" colourspace setting is only used for the selection of colours, not its mapping.

There are a number common reasons, usually because you need more control of the specific palette of colours used in an image. If you know of another reason to use the"-map" operator which I have not presented below - Mail me.

Common or'Best' Colourmap

The other technique, when handling multiple images, is to generate a common colour table for all the images involved.

Basically you append all the images together into one large image, then use the"**-colours**" operator to figure out a good colour map to use that is common to all the images. Once you have that colour map image you can use it to re-colour each of the original images using this juts generated Pre-Defined Colour Map.

Alternatively, you can use the special"+map" operator, which does the same thing to a 255 colour colourmap. It counts up the colours, perform the colour quantization to form a good, common colourmap, then dithers the images to use that map, if needed.

Both"**-map**" and"**+map**" forms however has one very important feature for GIF animations. It converts all the images to a image"**-type**" of'**Palette**' with all the images using the same palette of colours. The reason is that when writing a GIF image, the first images colour palette, will be used for the file formats'global colourmap'. Then as each image is written it notes that those images use the same set of colours, so it does NOT create a'local colourmap'. This can save up to 256 × 3, or 768 bytes of colourmap space for each and every image the final GIF file.

Only the"**-map**" operators can do this. So when handling GIF's and in particular GIF animations, it can be an important point to remember.

Web Safe Colouring

When the WWW was first created, computer displays had a limited range of colours available, and web browsers usually used a simpler set of colours for images. As such it was common to recolour images to this colour set, to make them both smaller, and to ensure they will look okay on users browsers.

To help with this IM provided a built-in colourmap image of this special table of 216 colours, called"**netscape:**". So lets look at how our test image would look on of a old web browser display using these colours.

```
convert logo.png -map netscape: map_netscape.gif
convert logo.png +dither -map netscape: map_netscape_nd.gif
```

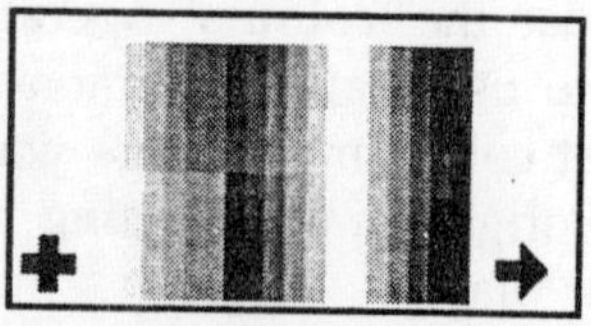

netscape:" colour map. This provides a set of 216 colours, which will nicely fit into the GIF formats 256 colour limit, and still have space for handling image transparency, or even adding some extracolours for special purposes, like shadows, or text overlays.

This colour map was generated by creating 6 levels of colours for each of the three colour channels, producing 6×6×6 colours or 216 colours. The number of the beast.

As only 219 colour are used it still has space (for GIF images) to add more colours to the colour map for specific purposes. For example a transparent colour, as well as more grey scale shades. An old Macintosh version of the web-safe map actually did exactly this to try to improve its overall result, but it was only used on Macintosh web clients.

Uniform 332 Colourmap

Another uniform colour mapping that is commonly used is the"332 RGB colour map". The number refer to the number of bits used to represent each colour within a 8 bit colour index. That is 3 bits (or 8 levels) of red, 3 for green, and 2 bits (or 4 colour levels) for blue, seeing as our eyes do not respond well to blue.

This gives 3+3+2 bits or an 8 bit colour index, or 256 colours. Perfect for the limited GIF colour table. However it will not leave any space for a GIF transparency colour, or other special use colours.

Here is one way to get IM to generate this colour map:

```
 convert -size 16x16 xc: -channel R -fx'(i%8)/
7' \
 -channel G -fx'(j%8)/7' \
 -channel B -fx'((i>>3&1)|(j>>2&2))/3' \
 -scale 600% colourmap_332.png
```

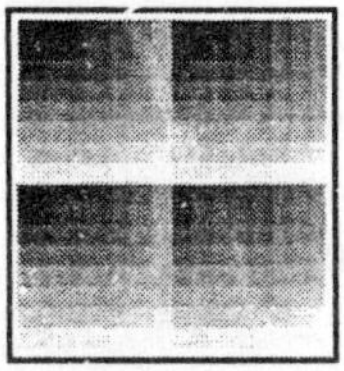

The only drawback with this map is that it does not actually provide any'gray' colours at all. However this drawback can be a plus when dithering is used as the difference produces a less distinct colour boundary changes in a grey-scale gradient, making it just that little bit smother looking.

TrueColour 16bit Colourmap

A similar uniform colourmap to the'332 colourmap' above is used by X windows in a rarely used 16 bit visual class. In this case, 16 bits are used for the colour index which is divided into 5 bits for red, 5 for green, and 6 for blue.

In other words, this colour map is more like a"556 colourmap". It is however rarely seen outside this environment, as images usually use an 8 bit colour table or 8 bit colour levels (8 bits per RGB value, or 24 bits, 32 with transparency).

Gamma Corrected Uniform Colourmaps

```
FUTURE: gamma-corrected uniform palette
http://www.compupter.com/uni.htm
Figure out gamma corrected colourmaps, 6 level,
332,
as well as the full 256 colourmap defined by
this article.
Figure out how to incorporate these into the -
ordered-dither expansion.
```

Posterize, Recolour using a Uniform Colour Map

The operators original purpose (using an argument of'2') is to re-colour images using just 8 basic colours, as if the image was generated using a simple and cheap poster printing method using just the basic colours. Thus the operator gets its name.

The"**-posterize**" operator is actual fact is a special colour reduction operator that generates a colour map based on the the number of colour'levels' given, for each colour channels in the image, dithering the image using an error correction dither.

```
convert netscape: -scale 50% +dither -posterize
2 posterize_2_ns.gif
convert netscape: -scale 50% +dither -posterize
3 posterize_3_ns.gif
convert netscape: -scale 50% +dither -posterize
6 posterize_6_ns.gif
```

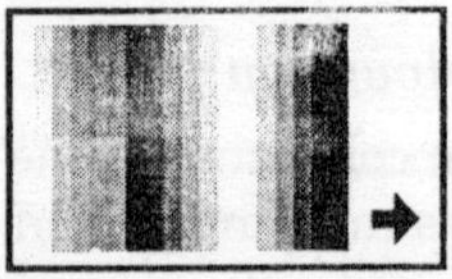

-posterize" argument on'2' means to only provide 2 colours per colour channel, producing a map of just 8 colours for a 3 channel RGB image, such as the above. Basically it will recolour images using the threshold set of 8 colours.

An argument of'**3**' will map image colours based on a colourmap of 27 colours, including mid-tone colours. While an argument of'**4**' will generate a 64 colour colourtable, and'5' generates a 125 colour colourmap. Of course as mentioned above, an argument of'6' will reproduce the same set of 216 colours, as provided by the in the built-in"**netscape:**" image.

Please note that a"**-posterize**" argument of'**0**' or'**1**' is non-sensible, and with the latest IM releases just converts images to pure black (which though logical, is quite useless). The result is that the image has been recolours using a mathematically derived or'uniform' colour map. You can see this clearing on a gradient image, producing an even distribution of posterized gray levels.

```
#convert -size 20x640 gradient: -rotate 90
gradient.png
convert gradient.png +dither -posterize 5
posterize_gradient.gif
```

As a better test lets poserize the shaded"colourwheel" image:

```
convert colourwheel.png +dither -posterize 2
posterize_2_cw.gif
convert colourwheel.png +dither -posterize 3
posterize_3_cw.gif
convert colourwheel.png +dither -posterize 6
posterize_6_cw.gif
```

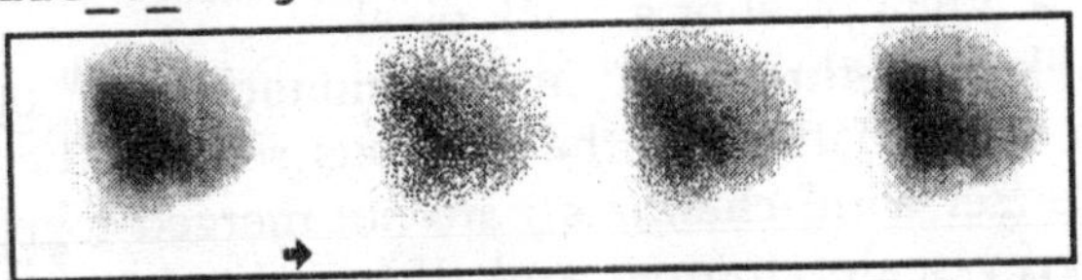

Of course many of the bitmap dithers we look at in the next section can also generate level 2 ordered dithers, using various sorts of dither styles. However few can use a larger number of grey levels.

Ordered Dither as of IM v6.2.9, is also a posterization method, due to its current limitation in dithering using a uniform colourmaps. However the dither pattern is more uniform, with a larger selection of styles, than the pseudo-randomized dither produced by"-posterize". Compare these with the dithered"-posterize" versions above.

```
convert colourwheel.png -ordered-dither o8x8,2
posterize_2_od.gif
convert colourwheel.png -ordered-dither o8x8,3
posterize_3_od.gif
convert colourwheel.png -ordered-dither o8x8,6
posterize_6_od.gif
```

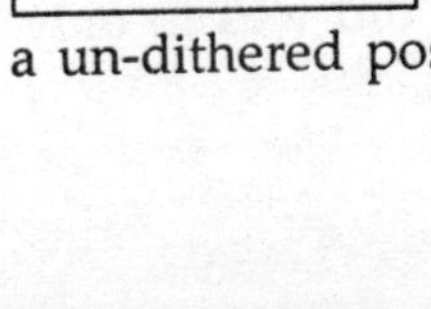

-ordered-dither" into a un-dithered posterization method.

Finally the Ordered Dither does allow you to specify a different number of colour levels, for each individual colour channel. Something that the"-posterize" operator does not currently allow.

Threshold Dithering Methods

Random Dither and Threshold

The"**-random-threshold**" operator is a special form of bitmap image converter. In this case it uses a very simple"random dither" to determine if a particular pixel is to become a white pixel or a black pixel.

Unlike the"**-threshold**" or the"**-monochrome**" operators, or even the variations in the previous section, the selected channels (set with"**-channels**") are not merged together into single grey-scale channel and dithered as a single unit. Instead"-random-threshold" works on each selected channel completely independently of each other.

Of course using the operator directly will result in a 2 level posterization of the image using a random dither.

```
convert logo.png -random-threshold 0x100%
random_posterize.gif
```

Converting to gray-scale will equalize all the channels in the image, before they are dithered. But as each channel is dithered independent of each other and in a random way, the result is not a bitmap image as you would expect. Instead you will get a splatter of colour pixels, especially for mid-tone colours.

```
convert logo.png -colourspace Gray -random-
threshold 0x100% \
 random_greyscale.gif
```

Here is the correct way to generate a proper randon dithered bitmap image.

```
convert logo.png -colourspace Gray -channel B
\
  -random-threshold 0x100% -separate
random_monochome.gif
```

Basically what it did was to only dither one channel of

the grey-scaled image, and then use the"-separate" channel operator to extract that channel as the final bitmap image.

As a special feature for this operator, IM will ensure a bitmap image is generated in the special"-channels" option of'All' is used.

```
convert logo.png -channel All -random-threshold
0x100% random_all.gif
```

Now that you know how to use the operator to correctly generate bitmaps from a colour image lets look how the argument effects the range of the dithering. This also shows clearly the'clumping' of pixels that this dither produces.

```
#convert -size 20x640 gradient: -rotate 90
gradient.png
convert gradient.png -channel All \
-random-threshold 0x100% random_grad_0x100.gif
convert gradient.png -channel All \
-random-threshold 10x90% random_grad_10x90.gif
convert gradient.png -channel All \
-random-threshold 25x75% random_grad_25x75.gif
convert gradient.png -channel All \
-random-threshold 50x50% random_grad_50x50.gif
```

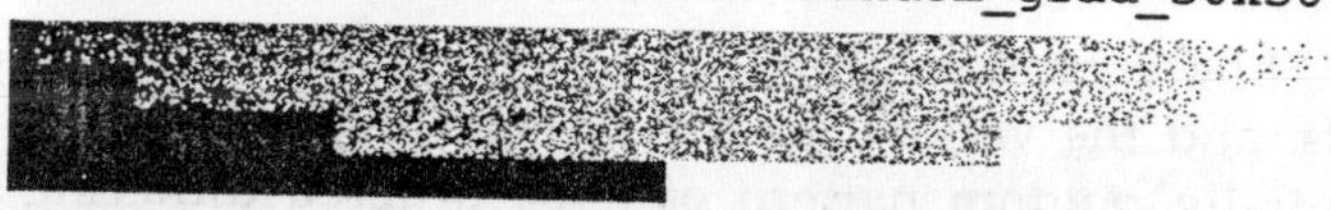

A"**-random-threshold**" setting of'**0x100%**' will produce a purely'Random Dither' of the image. If the two bounds are set to the same value (or even past each other) it will just produce a pure"**-threshold**" image.

Using any other sets of bounds (usually specified using a per centage) will threshold the bitmap outside the given range, while producing a random dither pattern for the values within the given range.

The best results can be obtained by using a slightly smaller range, just as you get using the"-monochrome" operator. A value of about'30x80%' probably the best result for most cases.

```
convert logo.png -channel All -random-threshold
30x80% random_30x80.gif
```

Of course the result is still not very good. But then this is the simplest and worst form of dithering you can get.

What actually happens is that the randomized dither pattern tends to produce'clumps' of pixels rather than a smooth dither pattern. this is due to the high frequency'noise' in the random number generator.

However at very high resolutions a random dither has been shown to produce a extremely good result, if random enough. IM uses a cryptographic level of randomness so it will likely be very random, though images are rarely used at a high enough a resolution to make it useful in this way.

One'fix' for this dither, that has been suggested is to use a random'blue noise' generator (a high frequency filter, as opposed to a low frequency'pink noise' filter used in sound production).

This should remove the clumping of the pixels, but is very difficult to implement digitally. No known implementation of'blue noise randomized dither' has been found, and unlikely to be ever created.

Ordered Dithering

While a random dither produces random clumps of pixels, and the various error correction dithers produce a essentially random pattern of dots, ordered dithering is basically the opposite. It is designed to be as mathematically deterministic as possible. So deterministic, that you actually need to specify the pattern for it to should use in dithering the image.

The"**-ordered-dither**" operator will dither each of the selected"**-channels**" in the image a given predefined pattern. The argument defines the pattern (known as a threshold map) to use.

These threshold maps fall into three basic styles. Diffused Pixel Dithers where the pixels are placed as far from each other as possible, so as to avoid'clumping' and tiling artifacts. Or to clump them together into tight dots that makes them easier to mechanically print, in a technique known as Digital Halftoning.

There is also some specialized artistic threshold maps that we will also be looking at, and even designing our own dithering pattern or threshold maps.

In each case the number of pixels that is on or off in the threshold map depends on the grey-level intensity of the image (or individual colour channel) that is being dithered into a bitmap.

The map adds pixel threshold levels in a consistent manner, so that once a pixel turns on at a particular'threshold' it remains on for any lighter grey colour. This consistency is very important, otherwise artifacts are produced along the boundaries of changes the dither pattern.

The important point about this is that the result for each pixel in an image is purely mathematically determined independently of any other pixel in the image. As such any small change to the original image will have absolutely no effect on the image in any other area a problem that Error Correction

FINAL OPTIMIZATIONS

There are some final things that can be used to gain more speed. Most of the time is spent finding the node which has to be reduced. It doesn't hurt to maintain a list of all nodes that are on the same level (level means how deep the node is stored in the tree).

A simple list could be very usefull for this. The best node to reduce is always stored at the highest level in the tree. It also doesn't hurt to reduce random nodes and don't analyse the contents of the node as long as you reduce in the highest level. The palette won't be as good, but it'll be a lot faster.

Also after you've created the palette the tree is still usefull. You can use them to fast map a RGB-tripple to a palette index.

To do this you have to store the palette indice into the nodes during the palette calculation. When you want to find a colour you simply traverse the tree down until you find a node which reference-count is greater than zero. The palette indice can be directly read out of the node. To my knowledge this is the fastest way to map RGB colours to adaptive palettes. Normally you don't need to use all 8 bits of the colour-components.

Using only 6 bits will give more performance and because standard VGA-adapters can only use 6 bits per primary colour noone would be able to spot the difference at all. It also saves a lot of memory

If memory-usage is an issue you might reduce the nodes after you inserted a complete scanline.

VECTOR QUANTIZATION

Vector quantization (VQ) is a lossy data compression method based on the principle of block coding. It is a fixed-to-fixed length algorithm.

In the earlier days, the design of a vector quantizer (VQ) is considered to be a challenging problem due to the need for multi-dimensional integration. In 1980, Linde, Buzo, and Gray (LBG) proposed a VQ design algorithm based on a training sequence.

The use of a training sequence bypasses the need for multi-dimensional integration. A VQ that is designed using this algorithm are referred to in the literature as an LBG-VQ.

Preliminaries

A VQ is nothing more than an approximator. The idea is similar to that of ``rounding-off'' (say to the nearest integer). An example of a 1-dimensional VQ is shown below:

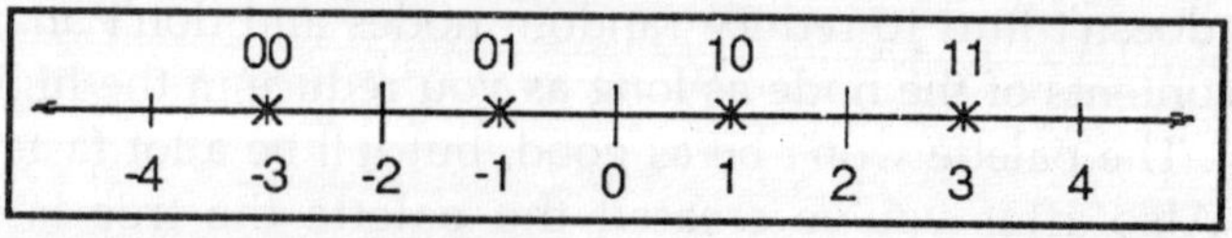

Here, every number less than -2 are approximated by -3. Every number between -2 and 0 are approximated by -1. Every number between 0 and 2 are approximated by +1. Every number greater than 2 are approximated by +3. Note that the approximate values are uniquely represented by 2 bits. This is a 1-dimensional, 2-bit VQ. It has a rate of 2 bits/dimension.

An example of a 2-dimensional VQ is shown below:

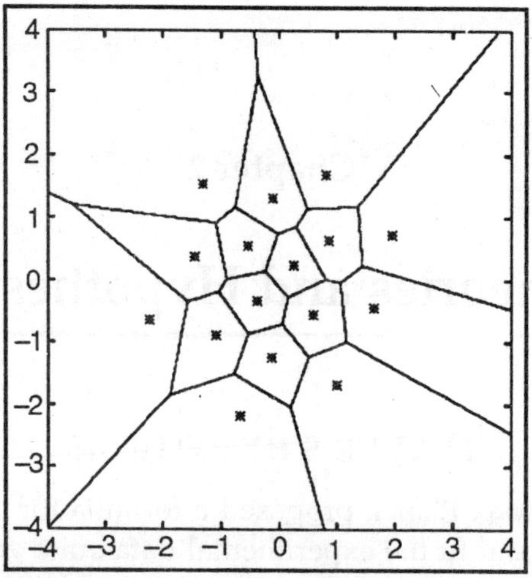

Here, every pair of numbers falling in a particular region are approximated by a red star associated with that region. Note that there are 16 regions and 16 red stars -- each of which can be uniquely represented by 4 bits. Thus, this is a 2-dimensional, 4-bit VQ. Its rate is also 2 bits/dimension.

In the above two examples, the red stars are called codevectors and the regions defined by the blue borders are called encoding regions. The set of all codevectors is called the codebook and the set of all encoding regions is called the partition of the space.

Chapter 2

Theories and Hypothesis

PLANCK'S HYPOTHESIS

In 1900 Max Planck proposed a formula for the intensity curve which did fit the experimental data quite well. He then set out to find a set of assumptions -- a model -- that would produce his formula. Instead of allowing energy to be continuously distributed among all frequencies, Planck's model required that the energy in the atomic vibrations of frequency f was some integer times a small, minimum, discrete energy,

$$E_{min} = hf$$

where h is a constant, now known as Planck's constant,

$$h = 6.626176 \times 10^{-34} \text{ J s}$$

Planck's proposal, then requires that all the energy in the atomic vibrations with frequency f can be written as,

$$E = n\,h\,f$$

where n in an integer, n = 1, 2, 3... No other values of the energy were allowed. The atomic oscillators could not have energy of (2.73) hf or (5/8) hf.

This idea that something -- the energy in this case -- can have only certain discrete values is called quantization. We say that the energy is quantized. This is referred to as Planck's quantum hypothesis."Quantum" means how great or of a fixed size. Planck did not realise how radical and far-reaching his proposals were. He viewed his strange assumptions as mathematical constructions to provide a formula that fit the experimental data. It was not until later, when Einstein used very similar ideas to explain the Photoelectric Effect in 1905, that it was realised that these assumptions described"real Physics" and were much more than mathematical constructions to provide the right formula.

In classical physics, matter is composed of tiny particles called atoms. It you assume this, it explain a lot of chemical and physical properties, the laws of chemical combination, the behaviour of gases, and so on, On the other hand, light in classical physics, is a phenomenon of continuous electromagnetic waves. How do we know that light is a wave phenomenon? We know because we can observe constructive and destructive interference of light waves in Thomas Young's two-slit experiment, which we actually performed, to see for ourselves that light indeed does travel as waves.

So now we're going to continue the story into the 20th century, where two revolutionary thinkers, Max Planck and Albert Einstein, began to question the 19th century synthesis and introduce quantum ideas into physics. Why would they do that? What was wrong with classical physics?

The problem was that there were just a few leftover experimental puzzles about light and matter. To solve them, they needed to change the entire structure of physics. The first puzzle was the problem of thermal radiation. When we heat up a solid object, it gives radiation in the form of long wavelength light. An everyday example of that is a light bulb. When we run an electric current through the tungsten filament of a light bulb, that filament gets very hot, so that it emits light.

How much radiation is emitted? What frequencics are emitted? They might be in thc visible range, that narrow range of frequencies we can perceive without eycs. There could be lower frequency light than that, the infrared light. There could be higher frequency light, the ultraviolet light.

When we Analyse the thermal radiation from an object, the simplest case actually turns out to be when the object itself is black in colour, whats called a blackbody. The reason is kind of paradoxical and surprising, since it turns out the more efficient an object is at absorbing light, the more efficient it is at emitting thermal radiation when you heat it up. So a blackbody would be a perfect radiator of thermal radiation.

It turns out that this thermal radiation, called blackbody radiation, is quite interesting. At a given temperature, all blackbodies, no matter what they're made of, radiate in exactly the same way. So at the end of the 19th century, many scientists tried to apply the classical theory of heat, to explain blackbody radiation. It turns out that this works pretty well for explaining the low frequency light, the infrared radiation that comes off of a blackbody.

Yet the classical theory predicts way too much high frequency radiation, way too much ultraviolet radiation. So back to the light bulb. We run the electric current through the tungsten filament, which gets very hot, more than 2000 degrees Celsius in fact.

Yet there's also a good deal of visible light, and there's almost no ultraviolet light from an incandescent light bulb. This is why you don't need to wear sunscreen around them, since they don't emit enough UV light to give you a sunburn. So lets try to apply the classical theory of heat to this. We then find out that UV light coming from the bulb, should in fact be much more intense than the visible light, which should in turn be much more intense than the infrared light. So we get it exactly wrong. There should be immense amounts of UV light coming out of the bulb.

This prediction is so radically wrong, that it came to be called the ultraviolet catastrophe. It your theory cannot explain how a light bulb works, you need a better theory! That's where Max Planck 1858-1947 came in. At the time of our story, Planck is a German physicist at age 42, quite old for a revolutionary in fact, and he's been working on the blackbody problem for some time. Planck realises that this is a really important problem, which will teach him something very fundamental

Why does he think this? It's because all blackbodies, whatever they are made of, radiate in exactly the same way. So if he can figure out why that is, he'll learn something fundamental about the relationship of heat and light. So he's been working on it for a long time, and finally in desperation, by 1900, he proposes a strange hypothesis. Planck says that suppose light is not emitted and absorbed continuously, but only in discrete amounts? These discrete amounts, he called light quanta, which is the quantum hypothesis, the quantum in quantum mechanics.

The energy in one light quantum, is related to the frequency of the light. The equation is:

$$E = hf$$

where h is a new fundamental constant of nature, later called Planck's constant. This has a tiny value, 6.6 times ten to the -34th joule seconds. That's some 33 zeros after the decimal point, before we got to the first six. So we multiply Planck's constant with the frequency of the light, and that tells us the size of the quantum of light energy.

Now since Planck's constant is so very small, light quanta are very small as well. They're very tiny, which is why energy appears to be so continuous. It comes in discrete units, yet they are so very small. The light bulb, under Planck's hypothesis, would emit something like ten to the 20th power, a hundred million trillion light quanta every second.

How does the light quanta hypothesis, explain blackbody radiation? Well, higher frequency light has quanta that have higher energy.

If an object is going to emit a quantum of light, it has to get together in one place, enough energy to do so. So it's just harder for an object to emit an ultraviolet photon. The UV photons have high frequency, higher energy, so it's just harder to emit them. So Planck's hypothesis predicts no ultraviolet catastrophe. The quantum hypothesis in fact, exactly accounts for all the details of blackbody radiation.

Now Planck's idea is pretty radical indeed. A continuous wave should carry any amount of energy, since the amount of that energy is just related to how strong the wave is, and you can make a continuous wave of any strength. Planck himself recognizes how radical his quantum hypothesis was. On the very day that we worked out that his quantum hypothesis solved the blackbody radiation problem, he said to his son:

"Today, I have made a discovery as important as Newton's discovery." That's quite a statement to make.

That brings us to the second puzzle, the problem of the photoelectric effect. Consider the following experiment. Suppose we take a metal surface and we polish it to remove any layer of oxides off it, and put it in a vacuum. Now we shine light onto the polished metal surface. What happens? It turns out that electrons, tiny electrically charged particles, constituents of atoms, are emitted from the surface. So this is why it's called the photoelectric effect,"photo" for the light, and"electric" because you emit electrons.

This was discovered by the late 1800s, and very soon, scientists had done experiments on it, making some curious observations about the nature of the photoelectric effect. They could measure both the number of electron emitted by the surface, and the energy of the electron, how fast they were going. They discovered that the energy of the electrons, how fast they were going, was not dependent at all on the intensity of the light that you shined. If you shine a brighter light, you get more electrons coming out of the surface, but every electron is moving at the same speed as previously. This is quite amazing.

If you imagine that the electrons are sort of being shaken out of the metal by the oscillations of the light, then if the oscillations are stronger, surely the electrons should be more shaken by the light? Yet even if you shined a light of a thousand times greater intensity on the metal surface, you won't have more energy in the electrons, you'll just have more electrons. That's very strange.

What does determine the energy of the electron then? It turns out that's dependent on the frequency of the light. If the frequency is too low in fact, you get no photoelectrons at all. Infrared light may not produce any photoelectrons from your metal surface. At higher frequency, you get electrons out, and at even high frequency, you get faster, higher energy electrons out. So the observation is that the energy of the electrons does not depend on light intensity, but only on the frequency of the light. That seems very puzzling indeed.

So along comes Albert Einstein, a very romantic figure in the history of science, being justly famous for all kinds of discoveries. This is the year of 1905, Einstein's year of miracles when he discovers Relativity theory, demonstrates $E=mc^2$, among all kinds of other things. Yet he also provides a new theory to explain the photoelectric effect.

By day in 1905, he's a technical expert third class in the Swiss patent office. By night, he's a physics revolutionary! Einstein is aware of Planck's work, so he realises that Planck's quantum hypothesis amounts to thinking of light as a stream of discrete particles, later called photons. So he realised, somehow that all meant light was made of photons.

So the energy of one photon, is determined by Planck's formula:

$$E=hf$$

That's sort of a strange picture, so how does it explain the photoelectric effect? Einstein says that every photoelectron, each and every one that comes out of the metal, gets its energy from a single photon striking the metal. Now some of the energy of the photon is used up in prying the electron loose from the metal, since the electrons are bound to the metal and it just takes a little energy to pry them loose. Yet whatever energy is left over afterward, the electron gets.

So photons in bright or dim light of a given colour, have the same energy. Yet if you change the colour, the frequency of the light, then the photons have different energies. The same intensity of light with higher frequencies, should have photons with higher energy, so the photoelectrons that are produced will also have a higher energy. So Einstein is then able to give a detailed account of the photoelectric effect in term of electrons. Now this is a very radical idea, and in fact, several experimental physicists were very skeptical. So they went off and did the experiments very carefully. They measured all kinds of things about the photoelectric effect, and they found the experimental results agreed exactly with Einstein's theory in every detail. So in fact, he won the Nobel Prize in 1921 for the photoelectric effect, not for Relativity Theory or any of the many other things he did, the photoelectric effect was that important.

So with Planck's explanation of blackbody radiation, and Einstein's explanation of the photoelectric effect, we sort of have ourselves in a weird situation for the theory of light. The two-slit experiment we did last lecture, seems to prove that light is a continuous wave. It seems to be a conclusive experiment.

On the other hand, when we do the photoelectric effect, that seems to prove that light is a stream of discrete photons. So there's something really funny going on.

Yet there's not just something funny about light, which brings us to the third puzzle, the problem of heat capacities. This is the heat energy needed to raise an object's temperature by one degree Celsius.

How much energy do we have to add to the object, so that its temperature will go up by one degree Celsius? If we have an object with a very high heat capacity, then it's hard to heat or cool, since it takes a lot of energy to put in to make it hotter, and a lot taken out to make it cooler.

On the other hand, if the object has a very low heat capacity, it's relatively easy to heat it or cool it. You don't have to add much energy to heat it up, and you don't have to extract much energy to cool it down. So heat capacity is a very significant quantity in the theory of heat, which was one of the aspects of classical 19th century physics.

There was a long-standing puzzle in the 19th century about pure solids, a solid made of one type of atom, of one particular element. So to explain this puzzle, we'll introduce a couple of examples. Platinum metal and diamond, which is made in carbon. We figure as long as we're doing thought experiments, we might just as well use the most expensive materials.

So classical heat theory of the 19th century, predicts that all pure solids should have about the same heat capacity, for the same number of atoms. This works where at a given temperature, all atoms should vibrate with the same energy on average. Heat energy is the energy of microscopic, random vibrations of the atoms.

Platinum atoms are very heavy, so vibrate very slowly

at a given temperature. Carbon atoms are much lighter atoms, where in diamond are held together by very stiff covalent bonds. So they vibrate much faster at a given temperature.

Yet the amount of energy in a platinum atoms or carbon atom, should be about the same at a given temperature. That means that the prediction is all substances made of one kind of atom, should have about the same heat capacity, for a given number of atoms.

So what do the experiments show? They show it's not quite true. So lets consider doing this experiment, at three different temperatures. We can imagine doing it at a very high temperature of 1000 degrees Celsius, hotter than any oven, then at an intermediate temperature, such as room temperature of 20 degrees Celsius, or at a very cold temperature like –200 degrees Celsius, about that of liquid Nitrogen.

So we can measure the heat capacities of materials at different temperatures, and when we do it at the high value, we find that both platinum and diamond behave about as expected. The classical theory actually predicts that all solid objects should have a heat capacity of about 25 in joule degrees Celsius per mole. So both metals have heat capacities of about that amount. Maybe they're not exactly right, but they are pretty close to that number.

Now we imagine the experiment at room temperature and find platinum to behave roughly as expected, yet diamond's heat capacity is much too low. It takes much too little energy to change the temperature of diamond.

Again, we find, if we do the experiment at liquid Nitrogen temperatures, that both metals have unexpectedly low heat capacities. That of diamond is very tiny. So this doesn't agree at all with classical physics.

Code:

```
Temp 1000 20 -200
Pt   25   24 16
C 22 4 0.1
```

Along comes Einstein again, and does something really radical, strange, and amazing. He applies quantum ideas, not to light, but to the vibrations of atoms. He says that suppose atomic vibration energy comes in discrete quanta of size E=hf. In other words, the Planck formula applies not just to energy and light waves, but also to the energy in atomic vibrations. This is the first application of quantum physics to matter. There's no light anywhere in the problem, so it's irrelevant to measuring the heat capacity of these materials.

This is a direct challenge to Newton's mechanics, the most successful theory of physics that had stood for more than 200 years for the gold standard of accurate and far-reaching theories. It's a direct challenge to Newton's mechanics, because a vibration atom in this mechanics, can have any energy whatsoever.

So what does Einstein say? Well consider these pure solids. At very high temperature there's a lot of heat energy for all the atoms to vibrate just as expected. Yet suppose we consider the same solid at low temperature. Then there's not enough heat energy for all the atoms to have at least one quantum of energy, because the atoms can have zero, one or two quanta, and so on.

So if there's not enough for everyone to have one quantum, that means a lot of the atoms will not be vibrating at all. So Einstein will say if a lot aren't even vibrating, the heat capacity of the material will be lower than expected. So at high temperatures, they should behave about as expected, as there is plenty of heat energy. At low temperatures, a lot of the atoms are fixed in place without any quanta of vibration energy and only a few are vibrating, so the heat capacity is lower.

So what's the difference between high and low temperatures? What counts as a low temperature for a given substance? It depends on the substance, on the speed of vibration of the substance, since it depends on how big the quantum of energy is, therefor the frequency of vibration. So for diamond, you have very high frequency vibrations. That means that the quantum of vibration energy is relatively high. That means that both 200 degrees below zero, and 20 degrees above zero, count as low temperatures for diamond. Even at room temperature, most atoms id diamond are not vibrating, donot have one quantum of vibration.

For platinum, on the other hand, those atoms are much more massive. The vibrations are much slower, and so room temperature of 20 degrees Celsius counts as high for the platinum vibrations. There is plenty of energy to have energy quanta for all the atoms. Yet 200 degrees below zero, that's low temperature, even for platinum.

Einstein's idea, with a few refinements, actually precisely explains the observed heat capacities of pure solids of all kinds at all temperatures. It's a remarkable agreement with experiment.

OK, what's happened in the first decade of the 20th century? Max Planck and Albert Einstein have introduced quantum ideas to explain a series of otherwise baffling observations, blackbody radiation in the case of Planck, the photoelectric effect and later the heat capacity of pure solids explained by Einstein.

All three of these were puzzles of classical physics, yet in each case, Planck and Einstein find that energy comes in discrete units, discrete quanta, the quantum in quantum physics. The size of the quantum is E=hf, where h is the same Plank's constant for everything, including radiation, the photoelectric effect, and the heat capacities of pure solids.

This is one of those really interesting moments in the history of science, where we're on the verge of a new point of view. You can tell there's a great new principle of physics in the offing. A principle that is universal, because the quantum hypothesis appears to apply both to light and to matter.

Now the universal quantum principle that applies to both light and matter, is a direct challenge to the classical physics that prevailed at the end of the 19th century, because in classical physics, energy can come in any amount, it's a continuous quantity. So this means that Plank's and Einstein's discoveries tell us that the universe is discrete in ways we never expected. That the old resolution of the universe into matter made of discrete things and light made of continuous, wavy stuff, that old distinction must not be quite right.

Well, next time of course, we're going to probe the other half of the mystery. Not only is nature discrete in ways we never expected, but it will turn out that nature is also continuous in ways that we never expected. So far, we've seen that light, electromagnetic waves, can also act as a stream of particles, of photons. Yet next time, we'll find out that particles of matter that we find out were completely discrete, can also act as waves. So every sort of stuff has a little bit of things, and every sort of thing has a little bit of stuff in it.

MAX PLANCK

Max Planck's hypothesis of a quantum of action and energy solved a problem in the distribution of radiant energy in a black body at any temperature. But Planck disliked the idea that physical quantities might be discrete and not continuous. Nevetheless, Planck's constanth lies at the heart of quantum mechanics.

Planck was convinced that determinism and strict causality were essential requirements for physical science and so must be true."Just as no physicist will in the last resort acknowledge the play of chance in human nature, so no physiologist will admit the play of chance in the absolute sense.

"the assumption of chance in inorganic nature is incompatible with the working principle of natural science."We must admit that the mind of each one of our greatest geniuses - Aristotle, Kant or Leonardo, Goethe or Beethoven, Dante or Shakespeare - even at the moment of its highest flights of thought or in the most profound inner workings of the soul, was subject to the causal fiat and was a instrument in the hands of an almighty law which governs the world."

Planck looked very closely at the problem of free will, and gave a rough version of the logical opposition between determinism and blind chance in the standard argument against free will.

"And here a question arises which seems to set a definite impassable limit to the principle of strict causality, at least in the spiritual sphere. This question is of such urgent human interest that I think it will be well if I treat it here before I come to a close. It is the question of the freedom of the human will. Our own consciousness tells us that our wills are free. And the information which that consciousness directly gives us is the last and highest exercise of our powers of understanding.

Let us ask for a moment whether the human will is free or whether it is determined in a strictly causal way. These two alternatives seem definitely to exclude one another. And as the former has obviously to be answered in the affirmative, so the assumption of a law of strict causality operating in the universe seems to be reduced to an absurdity in at least this one instance. In other words, if we assume the law of strict dynamic causality as existing throughout the universe, how can we logically exclude the human will from its operation ?

Many are the attempts that have been made to solve this dilemma. The purpose which in most cases they have set themselves has been to establish an exact limit beyond which the law of causality does not apply. Recent developments in physical science have come into play here, and the freedom of the human will has been put forward as offering logical grounds for the acceptance of only a statistical causality operative in the physical universe. As I have already stated on other occasions, I do not at all agree with this attitude. If we should accept it, then the logical result would be to reduce the human will to an organ which would be subject to the sway of mere blind chance. In my opinion the question of the human will has nothing whatsoever to do with the opposition between causal and statistical physics. Its importance is of a much more profound character and is entirely independent of any physical or biological hypothesis.

With many famous philosophers, that the solution of the problem lies in quite another sphere. On close examination, the above-stated alternative - Is the human will free or is it determined by a law of strict causality? - is based on an inadmissible logical disjunction. The two cases opposed here are not exclusive of one another. What then does it mean if we say that the human will is causally determined? It can only have one meaning, which is that every single act of the will, with all its motives, can be foreseen and predicted, naturally only by somebody who knows the human being in question, with all his spiritual and physical characteristics, and who sees directly and clearly through his conscious and sub conscious life. But this would mean that such a person would be endowed with absolutely clear-seeing spiritual powers of vision; in other words he would be endowed with divine vision.

Now, in the sight of God all men are equal. Even the most highly gifted geniuses, such as a Goethe or a Mozart, are but as primitive beings the thread of whose innermost thought and most finely spun feelings is like a chain of pearls unrolling in regular succession before His eye. This does not belittle the greatness of great men. But it would be a piece of stupid sacrilege on our part if we were to arrogate to ourselves the power of being able, on the basis of our own studies, to see as clearly as the eye of God sees and to understand as clearly as the Divine Spirit understands.

The profound depths of thought cannot be penetrated by the ordinary intellect. And when we say that spiritual happenings are determined, the statement eludes the possibility of proof. It is of a metaphysical character, just as the statement that there exists an outer world of reality. But the statement that spiritual happenings are determined is logically unassailable, and it plays a very important role in our pursuit of knowledge, because it forms the basis of every attempt to understand the connections between spiritual events. No biographer will attempt to solve the question of the motives that govern the acts of his hero by attributing these to mere chance. He will rather attribute his inability to the lack of source materials or he will admit that his own powers of spiritual penetration are not capable of reaching down into the depths of these motives. And in practical everyday life our attitude to our fellow beings is based on the assumption that their words and actions are determined by distinct causes, which lie in the individual nature itself or in the environment, even though we admit that the source of these causes cannot be discovered by ourselves.

What do we then mean when we say that the human will is free? That we are always given the chance of choosing between two alternatives when it comes to a question of taking a decision. And this statement is not in contradiction with what I have already said. It would be in contradiction only if a man could perfectly see through himself as the eye of God sees through him; for then, on the basis of the law of causality, he would foresee every action of his own will and thus his will would no longer be free. But that case is logically excluded; for the most penetrative eye cannot see itself, no more than a working instrument can work upon itself.

The object and subject of an act of knowing can never be identical; for we can speak of the act of knowing only when the object to be known is not influenced by the action of the subject who initiates and performs the act of knowing. Therefore the question as to whether the law of causality applies in this case or in that is in itself senseless if you apply it to the action of your own will, just as if somebody were to ask whether he could lift himself above himself or race beyond his shadow.

In principle every man can apply the law of causality to the happenings of the world around him, in the spiritual as well as in the physical order, according to the measure of his own intellectual powers; but he can do this only when he is sure that the act of applying the law of causality does not influence the happening itself. And therefore he cannot apply the law of causality to his own future thoughts or to the acts of his own will. These are the only objects which for the individual himself do not come within the force of the law of causality in such a way that he can understand its play upon them.

And these objects are his dearest and most intimate treasures. On the wise management of them depend the peace and happiness of his life. The law of causality cannot lay down any line of action for him and it cannot relieve him from the rule of moral responsibility for his own doings; for the sanction of moral responsibility comes to him from another law, which has nothing to do with the law of causality. His own conscience is the tribunal of that law of moral responsibility and there he will always hear its promptings and its sanctions when he is willing to listen.

It is a dangerous act of self-delusion if one attempts to get rid of an unpleasant moral obligation by claiming that human action is the inevitable result of an inexorable law of nature.

The human being who looks upon his own future as already determined by fate, or the nation that believes in a prophecy which states that its decline is inexorably decreed by a law of nature, only acknowledges a lack of will power to struggle and win through.

BOHR'S MODEL

BOHR'S MODEL OF HYDROGEN ATOM

Line Spectra

Lasers emit radiation which is composed of a single wavelength. However, most common sources of emitted radiation (i.e. the sun, a lightbulb) produce radiation containing many different wavelengths.

When the different wavelengths of radiation are separated from such a source a spectrum is produced. A rainbow represents the spectrum of wavelengths of light contained in the light emitted by the sun.

- Sun light passing through a prism (or raindrops) is separated into its component wavelengths
- Sunlight is made up of a continuous spectrum of wavelengths (from red to violet) - there are no gaps
  Not all radiation sources emit a continuous spectrum of wavelengths of light
- When high voltage is applied to a glass tube containing various gasses under low pressure different coloured light is emitted
  - Neon gas produces a red-orange glow
  - Sodium gas produces a yellow glow

- When such light is passed through a prism only a few wavelengths are present in the resulting spectra
  - These appear as lines separated by dark areas, and thus are called line spectra

When the spectrum emitted by hydrogen gas was passed through a prism and separated into its constituent wavelengths four lines appeared at characteristic wavelengths

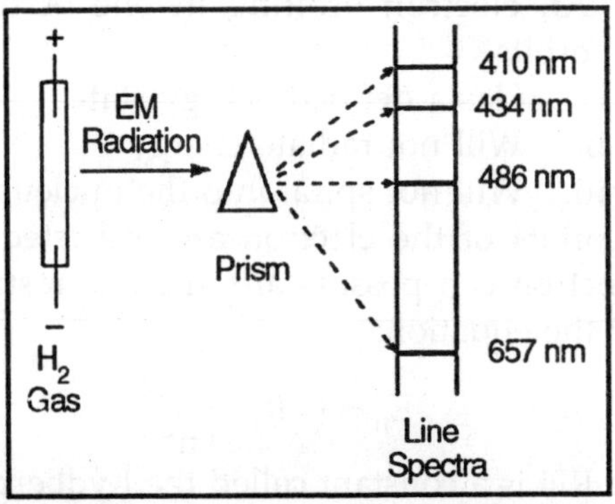

In 1885 a Swiss school teacher figured out that the frequencies of the light corresponding to these wavelengths fit a relatively simple mathematical formula:

$$v = C * \left(\frac{1}{2^2} - \frac{1}{n^2}\right) \text{where } n = 3,4,5,6$$

where C = 3.29 × 1015 s-1 (not the'c' used for the speed of light). However, the physical basis for this relationship was unknown.

BOHR'S RELATIONSHIP

- Bohr began with the assumption that electrons were orbiting the nucleus, much like the earth orbits the sun.
- From classical physics, a charge traveling in a circular path should lose energy by emitting electromagnetic radiation
- If the"orbiting" electron loses energy, it should end up spiraling into the nucleus (which it does not). Therefore, classical physical laws either don't apply

or are inadequate to explain the inner workings of the atom

- Bohr borrowed the idea of quantized energy from Planck
  - He proposed that only orbits of certain radii, corresponding to defined energies, are "permitted"
  - An electron orbiting in one of these"allowed" orbits.
    - i. Has a defined energy state
    - ii. Will not radiate energy
    - iii. Will not spiral into the nucleus

If the orbits of the electron are restricted, the energies that the electron can possess are likewise restricted and are defined by the equation:

$$E_n = (-R_H)\left(\frac{1}{n^2}\right)$$

Where RH is a constant called the Rydberg constant and has the value,

$$2.18 \times 10^{-18} \text{ J}$$

'n' is an integer, called the principle quantum number and corresponds to the different allowed orbits for the electron. Thus, an electron in the first allowed orbit (closest to the nucleus) has n=1, an electron in the next allowed orbit further from the nuclei has n=2, and so on.

Thus, the relative energies of these allowed orbits for the electrons can be diagrammed as follows:

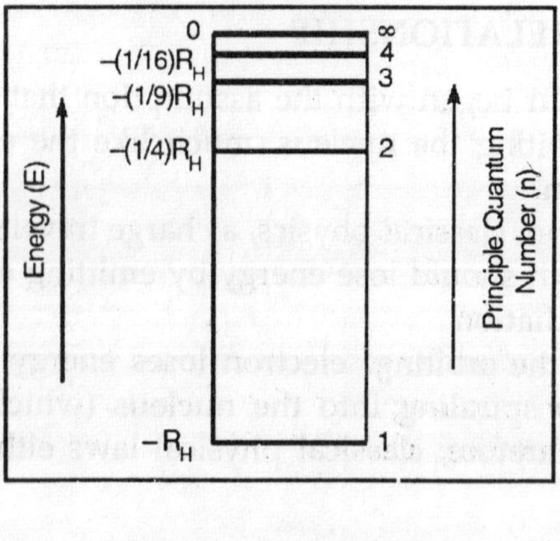

All the relative energies are negative:

- The lower the energy, the more stable the atom
- The lowest energy state (n=1) is called the ground state of the atom
- When an electron is in a higher (less negative) energy orbit (i.e. n=2 or higher) the atom is said to be in an excited state
- As n becomes larger, we reach a point at which the electron is completely separated from the nucleus
  - E = (– 2.18 × 10^{-18} J)(1/infinity) = 0
  - Thus, the state in which the electron is separated from the nucleus is the reference or zero energy state (actually higher in energy than other states)

Bohr also assumed that the electron can change from one allowed orbit to another:

- Energy must be absorbed for an electron to move to a higher state (one with a higher n value)
- Energy is emitted when the electron moves to an orbit of lower energy (one with a lower n value)
- The overall change in energy associated with"orbit jumping" is the difference in energy levels between the ending (final) and initial orbits:

$$\Delta E = E_f - E_i$$

Substituting in for the previously defined energy equation:

$$\Delta E = \left(\frac{-R_H}{n_f^2}\right) - \left(\frac{-R_H}{n_i^2}\right) = (-R_H) * \left(\frac{1}{n_f^2} - \frac{1}{n_i^2}\right) = R_H{}^* \left(\frac{1}{n_i^2} - \frac{1}{n_f^2}\right)$$

When an electron"falls" from a higher orbit to a lower one the energy difference is a defined amount and results in emitted electromagnetic radiation of a defined energy (ΔE)

- Planck had deduced that the energy of the photons comprising EM radiation is a function of its frequency (E = h)
- Therefore, if the emitted radiation from a falling electron had a defined energy, then it must have a correspondingly defined frequency

$$\Delta E = R_H{}^* \left(\frac{1}{n_i^2} - \frac{1}{n_f^2}\right) = hv$$

Note:

- ΔE is positive when NF is greater than Ni, this occurs when energy is absorbed and an electron moves up to a higher energy level (i.e. orbit).
- When E is negative, radiant energy is emitted and an electron has fallen down to a lower energy state

Revisiting Balmer's Equation

In 1885 a Swiss school teacher figured out that the frequencies of the light corresponding to these wavelengths fit a relatively simple mathematical formula:

$$v = C * \left(\frac{1}{2^2} - \frac{1}{n^2}\right) \text{ where } n = 3,4,5,6$$

where C = 3.29 × 1015 s-1 (not the'c' used for the speed of light) Since energy lost by the electrons is energy"gained" by the emitted EM energy, the EM energy from Bohr's equation would be:

$$hv = R_H * \left(\frac{1}{n_f^2} - \frac{1}{n_i^2}\right)$$

Thus, Balmer's constant'C' = (R_H/h) (Rydberg constant divided by Planck's constant), and nf = 2. Thus, the only emitted energies which fall in the visible spectrum are from those electrons which fell down to the second quantum orbital. Those which fell down to the first orbital have a higher energy (frequency) than can be seen in the visible spectrum.

Calculate the wavelength of light that corresponds to the transition of the electron from the n=4 to the n=2 state of the hydrogen atom. Is the light absorbed or emitted by the atom?

Since the electron is"falling" from level 4 down to level 2, energy will be given up and manifested as emitted electromagnetic radiation:

$$\Delta E = R_H * \left(\frac{1}{n_i^2} - \frac{1}{n_f^2}\right) = hv$$

ΔE = (2.18 × 10^{-18} J)((1/16)-(1/4)) = -4.09 × 10^{-19} J (light is emitted)

$4.09 \times 10\text{-}19$ J = $(6.63 \times 10^{-34}$ Js) * (v)

6.17×10^{14} s^{-1} =

$\lambda = (3.00 \times 10^{8}$ m $s^{-1})/ (6.17 \times 10^{14}$ $s^{-1}) = 4.87 \times 10^{-7}$m = 487 m

Bohr's model of the atom was important because it introduced quantized energy states for the electrons. However, as a model it was only useful for predicting the behaviour of atoms with a single electron (H, He+, and Li2+ ions). Thus, a different model of the atom eventually replaced Bohr's model. However, we will retain the concept of quantized energy states

BOHR MODEL OF ATOM

Structure of the Atom

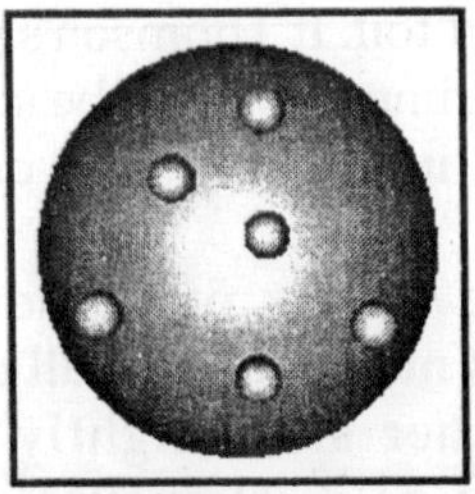

By the late nineteenth century, most people had accepted Dalton's proposal of 1808 that matter was made of atoms. Dalton"proved" his theory with a number of assumptions, each of which is either factually wrong or was used in a logically inconsistent manner.

Nonetheless opposition from critics such as Mach, who never believed in atoms, was largely ignored. Throughout the nineteenth century atomism became an idea that came to dominate thought in a number of fields, including political science, sociology, psychology, biology and more.

Then, J.J. Thomson discovered the electron. He determined that these electrons had a negative electric charge and compared to the atom had very little mass.

Thus he proposed that atoms consisted of a large massive positively charged body with a number of small negatively charged electrons scattered throughout it.

The total charge of the electrons exactly balanced the

positive charge of the large mass, so the total electric charge was zero. This was called the plum pudding model of the atom.

The number of electrons determines the particular chemical element. Hydrogen, for example, has one electron; helium has two; carbon has six, etc.

In the early twentieth century Rutherford was experimenting with one of the newly discovered radioactive substances, one that emitted alpha particles.

He knew that these particles had a mass much larger than the electron and had a net positive electric charge; now we know that these particles are identical to the nucleus of the helium atom.

He was directing a beam of these alpha particles onto a very thin piece of gold foil. If Thomson's plum pudding model was correct, the experiments would be sort of similar to firing BB's from a BB gun into a thin slab of cream cheese with chives.

And in fact the results of Rutherford's experiments usually followed this model: almost all of the alpha particles emerged on the other side slightly deflected by their interaction with the gold. However, once in a while he observed an alpha particle that was scattered right back towards the radioactive source.

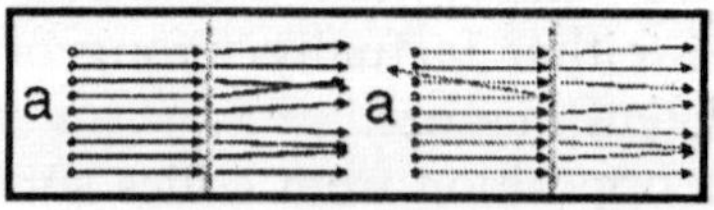

If you were doing the experiment involving the BB's and the cream cheese and occasionally had a BB scattered back towards you, you would probably conclude that there was something fairly small and very massive inside the slab. Similarly, Rutherford concluded that inside the gold foil there must be something fairly small, very massive, and positively charged.

Thus the plum pudding model of the atom collapsed: most of the mass and the positive charge of the atom was concentrated into a very small volume. This small massive positively charged object is called the nucleus.

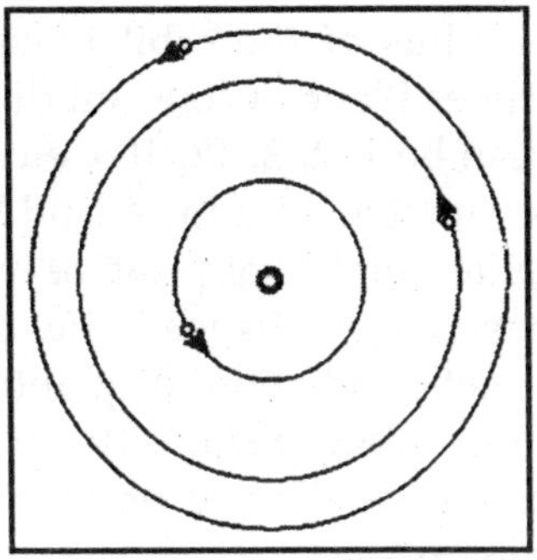

Soon, people proposed a planetary model of the atom. The electrons were in orbits around the nucleus, held in their orbits by the electric force that attracts negatively charged electrons to the positively charged nucleus.

However this model makes no sense. We know from classical electromagnetic theory that any charged body that is in a state of motion other than at rest or in uniform motion in a straight line will emit energy as electromagnetic radiation. Thus the electrons is this planetary model will be radiating energy. As they lose energy they will spiral into the nucleus and in a matter of nanoseconds will collide with it. Thus this atom can not be stable.

Neils Bohr knew about all of these facts, and in the early part of the century was collabourating with Rutherford. He also knew about the existence of line spectra from chemical elements; a document on this topic may be found here. He was struggling to make sense of all of this. As was common with Bohr when confronted with a puzzle, this struggle was nearly all-consuming.

Then in 1913 Bohr, by accident, stumbled across Balmer's numerology for the hydrogen spectrum, and in a flash came up with a workable model of the atom.

The model asserts that:

- The planetary model is correct.
- When an electron is in an"allowed" orbit it does not radiate.Thus the model simply throws out classical electromagnetic theory. Technical note: an allowed orbit is one in which the electron mass times its speed

times the radius of the orbit is equal to a positive integer n times Planck's constant divided by 2 pi. The integer n can be 1, 2, 3, 17, 108, etc. In fact, there are an infinite number of allowed orbits corresponding to the infinite number of positive integers.

- When an electron absorbs energy from incident electromagnetic radiation, it"quantum jumps" into a higher energy allowed state. This higher energy state corresponds to an allowed orbit with a higher value of the integer n.
- When an electron is in a higher energy state, it can quantum jump into a lower energy state, one with a smaller value of n, emitting all of its energy as a single photon of electromagnetic energy.

A Flash animation of Bohr's model showing the excitation and photon emission of the electron in a Hydrogen atom has been prepared. It requires the Flash player of Version 6 or better; the player is available free. Technical note: the allowed orbits are given by the equation to the right, where m is the mass of the electron, v is its speed, r is the radius of the orbit, and h is Planck's constant.

$$m v r = n \frac{h}{2\pi}$$

As Bohr fully realised, this model is largely ad hoc, if not downright ugly. It does, however,"explain" the line spectra of the elements. When Bohr published his model Otto Stern, who was Einstein's student, and Max von Laue, who was Planck's student, made an earnest vow:"If this nonsense of Bohr should in the end prove to be right, we will quit physics!"

Given these assertions plus the standard classical laws of attraction between two oppositely charged objects one can calculate the energy of the electrons in their orbits as a function of the value of n. It turns out to depend on the mass and charge of the electron, Planck's constant, some constants from electromagnetic theory, and also depends on 1/n2. The energy of allowed orbits for hydrogen is shown to the right; we have also indicated the value of n for three of those energy levels.

Technical note: the fact that the energies are negative is only due to a common convention as to where we choose the zero point of energy in a non-relativistic analysis.

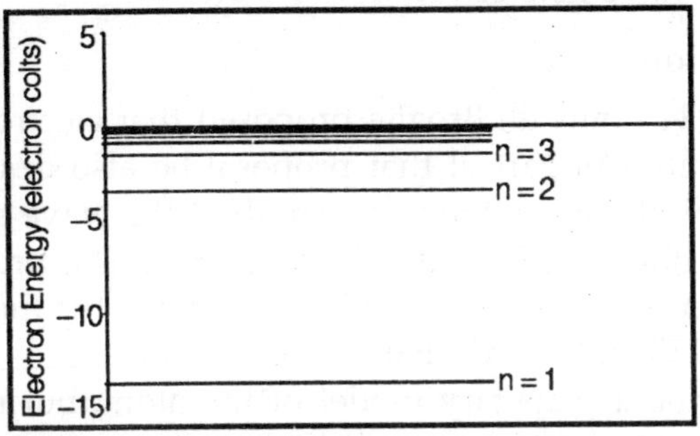

Recall Balmer's formula:

$$\frac{1}{\text{wavelength}} = R\left(\frac{1}{2^2} - \frac{1}{n^2}\right)$$

Here R is a constant equal to about 10,970,000 and n is any integer greater than 2, such as 3, 4, 5, 108, etc. Each different value of n gives the wavelength of a different line in the Hydrogen spectrum. The constant R is usually called theRydberg constant.

The wavelengths described by this formula correspond exactly to the radiation emitted when an electron in the Bohr model quantum jumps from a high energy orbit, described by some n greater than 2, to an orbit whose value of n is equal to 2. In the model transitions to other"final" states such as n equal to 1 or 3, 4, 5, etc. are also predicted and the wavelengths in the spectrum for these are found experimentally to exist. Further, the constant R turns out to be expressed in terms of fundamental physical constants. So we have pushed the numerology of Balmer's work a bit back.

Technical note: the Rydberg constant is given by the equation to the right, where m is the mass of the electron, e its charge, c is the speed of light, h is Planck's constant, and 60 the constant relating the forces exerted by charged bodies on each other and the magnitudes of the charges.

$$R = \frac{\pi^2 m e^4}{8 c h^3 60^2}$$

The lowest energy allowed orbit, the one with n equal to one, is the ground state of the atom. An unexcited hydrogen atom will have its electron in this state.

Explanation

In 1924 Louis de Broglie proposed that electrons have a wave nature. As part of that proposal he also described the relationship between the wavelength of the wave aspect and the mass and speed of its particle aspect. The proposal has been experimentally confirmed and is one of the fundamental aspects of Quantum Mechanics.

Imagine a planetary model of the atom, but now think of the electrons as waves. Which waves states might be allowed? You may recall that waves that"fit" into their available space, standing waves, are the ones that are allowed. A document discussing this is available.

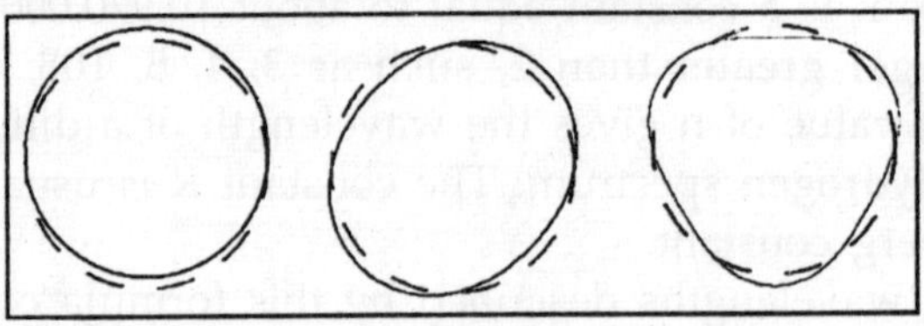

So in general the circumference equals n times the wavelength, where n is any positive integer. In the above figures the value of n is 1, 2 and 3 respectively.

You can view a simple animation of any of the three orbits above by pointing at it and clicking the left button. If you are using a fairly modern browser the animation will appear in a separate window; close that window whenever you wish. If you are using an older browser the animation will appear in this window; use the Back button of your browser to return here.

It turns out that these standing wave states for electrons correspond exactly to the"allowed" electron orbits in Bohr's model. So, Quantum Mechanics explains Bohr's ad-hoc model of the atom.

Now we call the integer n the principle quantum number. Note that for the hydrogen atom it completely describes the state of the electron.

EMISSION SPECTRUM OF HYDROGEN

When an electric current is passed through a glass tube that contains hydrogen gas at low pressure the tube gives off blue light. When this light is passed through a prism, four narrow bands of bright light are observed against a black background.

These narrow bands have the characteristic wavelengths and colours shown in the table below:

Wavelength	Colour
656.2	red
486.1	blue-green
434.0	blue-violet
410.1	violet

Four more series of lines were discovered in the emission spectrum of hydrogen by searching the infrared spectrum at longer wave-lengths and the ultraviolet spectrum at shorter wavelengths. Each of these lines fits the same general equation, where n_1 and n_2 are integers and R_H is 1.09678 × 10^{-2} nm^{-1}.

$$\frac{1}{\lambda} = R_H\left[\frac{1}{n_1^2} - \frac{1}{n_2^2}\right]$$

Explanation of Emission Spectrum

Max Planck presented a theoretical explanation of the spectrum of radiation emitted by an object that glows when heated. He argued that the walls of a glowing solid could be imagined to contain a series of resonators that oscillated at different frequencies. These resonators gain energy in the form of heat from the walls of the object and lose energy in the form of electromagnetic radiation. The energy of these resonators at any moment is proportional to the frequency with which they oscillate.

To fit the observed spectrum, Planck had to assume that the energy of these oscillators could take on only a limited number of values. In other words, the spectrum of energies for these oscillators was no longer continuous. Because the number of values of the energy of these oscillators is limited, they are theoretically"countable." The energy of the oscillators in this system is therefore said to be quantized. Planck introduced the notion of quantization to explain how light was emitted.

Albert Einstein extended Planck's work to the light that had been emitted. At a time when everyone agreed that light was a wave (and therefore continuous), Einstein suggested that it behaved as if it was a stream of small bundles, or packets, of energy. In other words, light was also quantized. Einstein's model was based on two assumptions. First, he assumed that light was composed of photons, which are small, discrete bundles of energy. Second, he assumed that the energy of a photon is proportional to its frequency.

$$E = hv$$

In this equation, h is a constant known as Planck's constant, which is equal to 6.626 × 10-34 J-s.

Example: Let's calculate the energy of a single photon of red light with a wavelength of 700.0 nm and the energy of a mole of these photons.

Red light with a wavelength of 700.0 nm has a frequency of 4.283 × 1014 s-1. Substituting this frequency into the Planck-Einstein equation gives the following result.

$$E = \left(6.626 \times 10^{-34}\,J - s\right)\left(4.283 \times 10^{14}\,s^{-1}\right) = 2.838 \times 10^{-19}\,J$$

A single photon of red light carries an insignificant amount of energy. But a mole of these photons carries about 171,000 joules of energy, or 171 kJ/mol.

$$\frac{2.838 \times 10^{-19}\,J}{1\text{ photon}} \times \frac{6.022 \times 10^{23}\text{ photons}}{1\text{ mole}} = 170.9\,kJ/mol$$

Absorption of a mole of photons of red light would therefore provide enough energy to raise the temperature of a liter of water by more than 40oC.

The fact that hydrogen atoms emit or absorb radiation at a limited number of frequencies implies that these atoms can only absorb radiation with certain energies. This suggests that there are only a limited number of energy levels within the hydrogen atom. These energy levels are countable. The energy levels of the hydrogen atom are quantized.

Proposal of Bohr Model

Niels Bohr proposed a model for the hydrogen atom that explained the spectrum of the hydrogen atom. The Bohr modelwas based on the following assumptions.

- The electron in a hydrogen atom travels around the nucleus in a circular orbit.
- The energy of the electron in an orbit is proportional to its distance from the nucleus. The further the electron is from the nucleus, the more energy it has.
- Only a limited number of orbits with certain energies are allowed. In other words, the orbits are quantized.
- The only orbits that are allowed are those for which the angular momentum of the electron is an integral multiple of Planck's constant divided by 2p.
- Light is absorbed when an electron jumps to a higher energy orbit and emitted when an electron falls into a lower energy orbit.
- The energy of the light emitted or absorbed is exactly equal to the difference between the energies of the orbits.

Some of the key elements of this hypothesis are illustrated in the figure below:

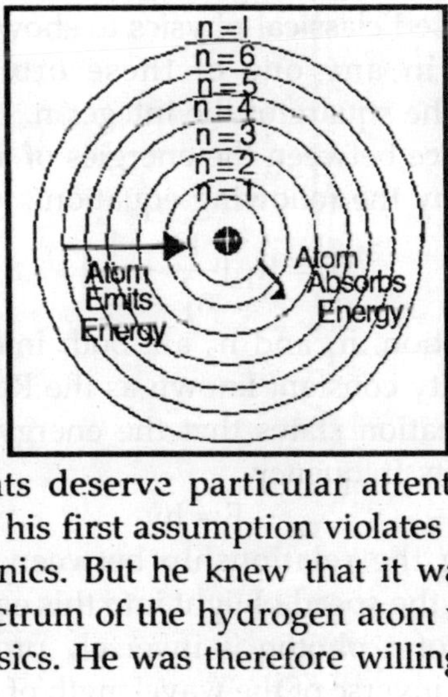

Three points deserve particular attention. First, Bohr recognized that his first assumption violates the principles of classical mechanics. But he knew that it was impossible to explain the spectrum of the hydrogen atom within the limits of classical physics. He was therefore willing to assume that one or more of the principles from classical physics might not be valid on the atomic scale.

Second, he assumed there are only a limited number of orbits in which the electron can reside. He based this assumption on the fact that there are only a limited number of lines in the spectrum of the hydrogen atom and his belief that these lines were the result of light being emitted or absorbed as an electron moved from one orbit to another in the atom.

Finally, Bohr restricted the number of orbits on the hydrogen atom by limiting the allowed values of the angular momentum of the electron.

Any object moving along a straight line has a momentum equal to the product of its mass (m) times the velocity (v) with which it moves. An object moving in a circular orbit has an angular momentum equal to its mass (m) times the velocity (v) times the radius of the orbit (r). Bohr assumed that the angular momentum of the electron can take on only certain values, equal to an integer times Planck's constant divided by 2p.

$$mvr = n[\frac{h}{2\pi}](\text{where } n = 1,2,3,4,...)$$

Bohr then used classical physics to show that the energy of an electron in any one of these orbits is inversely proportional to the square of the integer n.

The difference between the energies of any two orbits is therefore given by the following equation.

$$\Delta E = R_H[\frac{1}{n_1^2} - \frac{1}{n_2^2}]$$

In this equation, n_1 and n_2 are both integers and R_H is the proportionality constant known as the Rydberg constant.

Planck's equation states that the energy of a photon is proportional to its frequency.

$$E = hv$$

Substituting the relationship between the frequency, wavelength, and the speed of light into this equation suggests that the energy of a photon is inversely proportional to its wavelength. The inverse of the wavelength of electromagnetic radiation is therefore directly proportional to the energy of this radiation.

By properly defining the units of the constant, R_H, Bohr was able to show that the wavelengths of the light given off or absorbed by a hydrogen atom should be given by the following equation.

$$\frac{1}{\lambda} = R_H[\frac{1}{n_1^2} - \frac{1}{n_2^2}]$$

Thus, once he introduced his basic assumptions, Bohr was able to derive an equation that matched the relationship obtained from the analysis of the spectrum of the hydrogen atom.

According to the Bohr model, the wavelength of the light emitted by a hydrogen atom when the electron falls from a high energy (n = 4) orbit into a lower energy (n = 2) orbit.

Substituting the appropriate values of R_H, n_1, and n_2 into the equation shown above gives the following result.

$$\frac{1}{\lambda} = \left(1.09678 \times 10^{-2} \text{nm}^{-1}\right)[\frac{1}{2_1^2} - \frac{1}{4_2^2}]$$

Solving for the wavelength of this light gives a value of 486.3 nm, which agrees with the experimental value of 486.1 nm for the blue line in the visible spectrum of the hydrogen atom.

Wave-Particle Duality

The theory of wave-particle duality developed by Louis-Victor de Broglie eventually explained why the Bohr model was successful with atoms or ions that contained one electron. It also provided a basis for understanding why this model failed for more complex systems. Light acts as both a particle and a wave. In many ways light acts as a wave, with a characteristic frequency, wavelength, and amplitude. Light carries energy as if it contains discrete photons or packets of energy.

When an object behaves as a particle in motion, it has an energy proportional to its mass (m) and the speed with which it moves through space (s).

$$E = ms2$$

When it behaves as a wave, however, it has an energy that is proportional to its frequency:

$$E = hv = \frac{hs}{\lambda}$$

By simultaneously assuming that an object can be both a particle and a wave, de Broglie set up the following equation.

$$ms^2 = \frac{hs}{\lambda}$$

By rearranging this equation, he derived a relationship between one of the wave-like properties of matter and one of its properties as a particle.

As noted in the previous section, the product of the mass of an object times the speed with which it moves is themomentum (p) of the particle. Thus, the de Broglie equation suggests that the wavelength (l) of any object in motion is inversely proportional to its momentum.

$$\lambda = \frac{h}{p}$$

De Broglie concluded that most particles are too heavy to observe their wave properties. When the mass of an object is very small, however, the wave properties can be detected experimentally. De Broglie predicted that the mass of an electron was small enough to exhibit the properties of both particles and waves. In 1927 this prediction was confirmed when the diffraction of electrons was observed experimentally by C. J. Davisson.

The Bohr Model vs. Reality

At first glance, the Bohr model looks like a two-dimensional model of the atom because it restricts the motion of the electron to a circular orbit in a two-dimensional plane. In reality the Bohr model is a one-dimensional model, because a circle can be defined by specifying only one dimension: its radius, r. As a result, only one coordinate (n) is needed to describe the orbits in the Bohr model.

Unfortunately, electrons aren't particles that can be restricted to a one-dimensional circular orbit. They act to some extent as waves and therefore exist in three-dimensional space. The Bohr model works for one-electron atoms or ions only because certain factors present in more complex atoms are not present in these atoms or ions. To construct a model that describes the distribution of electrons in atoms that contain more than one electron we have to allow the electrons to occupy three-dimensional space. We therefore need a model that uses three coordinates to describe the distribution of electrons in these atoms.

Wave Functions and Orbitals

We still talk about the Bohr model of the atom even if the only thing this model can do is explain the spectrum of the hydrogen atom because it was the last model of the atom for which a simple physical picture can be constructed. It is easy to imagine an atom that consists of solid electrons revolving around the nucleus in circular orbits.

Erwin Schrödinger combined the equations for the behaviour of waves with the de Broglie equation to generate a mathematical model for the distribution of electrons in an atom. The advantage of this model is that it consists of mathematical equations known as wave functions that satisfy the requirements placed on the behaviour of electrons. The disadvantage is that it is difficult to imagine a physical model of electrons as waves.

The Schrödinger model assumes that the electron is a wave and tries to describe the regions in space, or orbitals, where electrons are most likely to be found. Instead of trying to tell us where the electron is at any time, the Schrödinger model describes the probability that an electron can be found in a given region of space at a given time. This model no longer tells us where the electron is; it only tells us where it might be.

BOHR AND THE QUANTUM ATOM

So far we have seen that quantum physics deals with electromagnetic radiation -- that is to say, light. But at the beginning we said that quantum physics tells us that material things are described by quantum physics. So what happened to matter?

Niels Bohr happened to matter. Bohr was a Danish physicist whose parents were both scientists. Apparently young Niels grew up in an atmosphere that was favourable to science. He received his doctorate in 1911 from the University of Copenhagen.

Niels Bohr

One can see just by looking at Niels Bohr's early academic life (though there are many other indicators) that the'atomic theory' of matter -- that is to say that matter is made up of tiny atoms -- had gone from being scientifically marginalised to being at the heart of physics in a very short time. Bohr's doctoral thesis was concerned with the inadequacy of classical (that is to say, Newtonian) physics for describing the behaviour of matter at the atomic level.

In the 19th century, physicists who saw some value in the atomic theory thought of the atom as a tiny undivided and indivisible unit of matter, the smallest possible unit into which something could be broken down.

Experiments dealing with the photoelectric effect, along with other observations, strongly suggested that the atom had some internal structure, since particles called'electrons' were being emitted from them.

RUTHERFORD'S MODEL

A model of the atom was described by the British physicist Ernest Rutherford in 1911, and is known as the Solar System model. It is very simple, and is still used to teach elementary atomic structure to school children.

- An atom consists of a central nucleus. This nucleus is composed of positively charged protons, and electrically uncharged (neutral) neutrons.
- Negatively charged electrons revolve round the nucleus in definite orbits.
- The orbits themselves can be at any distance from the nucleus.

- In any atom, the number of protons is equal to the number of electrons, and hence it is electrically neutral.

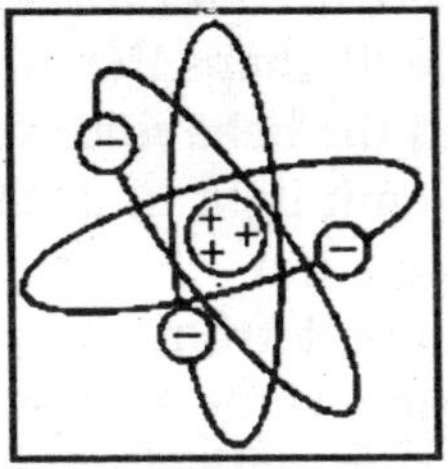

Fig. Rutherford's Atomic Model [1]

Rutherford used the laws of motion that had been set forward by Sir Isaac Newton to describe the atom. According to Rutherford's description, the electrons of an atom could occupy one of an infinite number of orbits, in accordance with Newton's laws. There were problems with Rutherford's description of the atom from the beginning. Let us find out two drawbacks of Rutherford's theory.

Inherent Instability of the Atom

According to Rutherford's theory, electrons could orbit the nucleus at any distance. When the electrons circle round the nucleus, they are constantly changing their direction. According to classical electrodynamics (which deals with the motion of electrons), such electrons which either constantly change their direction or their velocity or both should continuously emit radiation. While doing so, they should lose energy, and thus spiral into the nucleus. This means every atom is unstable, quite contrary to our observation!

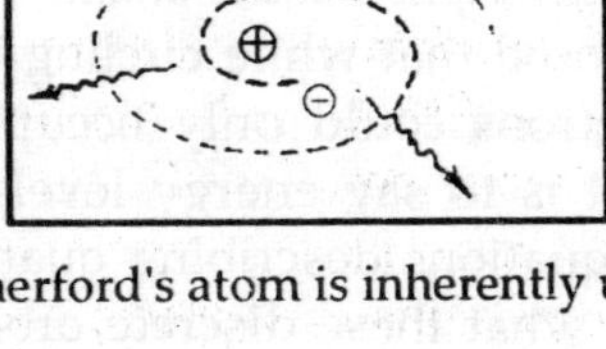

Fig. Rutherford's atom is inherently unstable [2]

Atomic Spectra

Rutherford's description of the atom could not be entirely correct because it did not account for some observations that had already been made. Perhaps the most important of these observations concerned the behaviour of certain gases. These gases at low pressure emit light in a set of discrete bands of the electromagnetic spectrum.

This is quite different from the radiation emitted by solids, which is spread evenly across the electromagnetic spectrum. The radiation emissions of these gases were important because they showed that at least under some circumstances, the orbits of the electrons could not be at just any distance from the nucleus, but were confined to discrete distances (or energy states).

Fig. A. Continuous Spectrum and B. line Spectrum of Hydrogen [3]

If the electrons in these gases were free to orbit at any distance, then the light emitted from them would have been spread evenly across the electromagnetic spectrum. Instead, what experimenters saw was that the light from these gases showed a distinct line pattern. That is to say that the light being emitted was only seen in a certain set of wavelengths, with empty spaces in between.

These line-spectra were different for each gas, and was found to be the characteristic of its atom. Today, astronomers use line-spectra to detect the elements present in stars.

Bohr's Explanation

Niels Bohr quickly seized upon this problem and used it to propose a quantised description of the atom:

- Bohr proposed that while circling the nucleus of the atom, electrons could only occupy certain discrete orbits, that is to say energy levels. Bohr used Max Planck's equations describing quanta of radiation to determine what these discrete orbits would have to be. As long as electrons stay in these energy levels, they are stable.

- Further, Bohr said electrons give or take energy only when they change their energy levels. If they move up, theytake energy (say from light), and if they move down, they release energy. This energy itself is released in discrete packets called photons, which were introduced.
- Furthermore, Bohr also said that an electron which is not in its native energy level (in other words, which has beenexcited to a higher energy level) always has to fall back to its original, stable level.

Bohr interpreted the lines in the spectra of gases as formed by the transitions of electrons to and from various energy-levels. This has been verified thoroughly with the hydrogen atom, and found to be correct. Bohr's formulae agreed excellently with observed line positions.

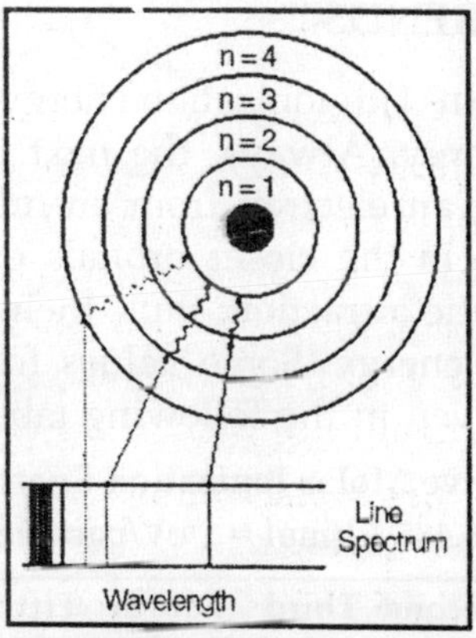

Fig. Bohr's Explanation of Line Spectra [4]

Imagine that you are taking a walking along a beach. As you walk along, you see a sand-castle that someone has built. As you get closer to the sand castle, you discover that you can only stand three metres, two metres, or one metre from the sand castle. You cannot stand at one and a half metres, nor can you stand at two and three quarters of a metre from the sand castle. No matter how hard you try, some mysterious force keeps you at one of those three distances. In everyday life such a situation is absurdly impossible. In the physics of the very small, it is a necessity.

This description in which electrons can only occupy certain orbits is called the shell model of the atom, because Bohr described the possible orbits of the electrons as orbitals or shells. When an atom of a gas released energy, an electron would move down to a lower orbit (requiring less energy), and when an atom acquired energy, an electron would moveup to a higher energy level. But these orbits or shells were discrete, like the distances from the sand castle. The orbits were not a smooth, continuous series of possibilities as one finds in the everyday world, but rather a set of distinct statesseparated from each other, much like the separation of the quanta of electromagnetic radiation that Planck had discovered. This caused the distinct lines in the spectrum.

For the first time, quantum physics had been applied to matter.

VALUES AND TRENDS

Generally the (n+1)th ionization energy is larger than the nth ionization energy. Always, the next ionization energy involves removing an electron from an orbital closer to the nucleus. Electrons in the closer orbitals experience greater forces of electrostatic attraction; thus, their removal requires increasingly more energy. Some values for elements of the third period are given in the following table:

Table. Successive Molar Ionization Energies in kJ/mol (96.485 kJ/mol = 1 eV/particle)

Element	First	Second	Third	Fourth	Fifth	Sixth	Seventh
Na	496	4,560					
Mg	738	1,450	7,730				
Al	577	1,816	2,881	11,600			
Si	786	1,577	3,228	4,354	16,100		
P	1,060	1,890	2,905	4,950	6,270	21,200	
S	999.6	2,260	3,375	4,565	6,950	8,490	27,107
Cl	1,256	2,295	3,850	5,160	6,560	9,360	11,000
Ar	1,520	2,665	3,945	5,770	7,230	8,780	12,000

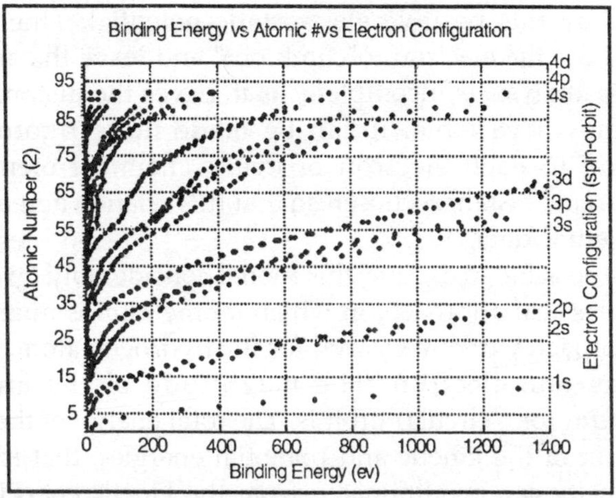

Large jumps in the successive molar ionization energies occur when passingnoble gas configurations. For example, as can be seen in the table above, the first two molar ionization energies of magnesium (stripping the two 3s electrons from a magnesium atom) are much smaller than the third, which requires stripping off a 2p electron from the very stable neon configuration of Mg^{2+}. Ionization energy is also a periodic trend within theperiodic table organization. Moving left to right within a period or upward within a group, the ionization energy generally increases. As theatomic radius decreases, it becomes harder to remove an electron that is closer to a more positively charged nucleus.

ELECTROSTATIC EXPLANATION

Atomic ionization energy can be predicted by an analysis using electrostatic potential and the Bohr model of the atom, as follows.

Consider an electron of charge $-e$ and an atomic nucleus with charge +Ze, where Z is the number of protons in the nucleus. According to the Bohr model, if the electron were to approach and bind with the atom, it would come to rest at a certain radius a. The electrostatic potential V at distance a from the ionic nucleus, referenced to a point infinitely far away, is: Since the electron is negatively charged, it is drawn inwards by this positive electrostatic potential. The energy required for the electron to"climb out" and leave the atom is:

This analysis is incomplete, as it leaves the distance a as an unknown variable. It can be made more rigorous by assigning to each electron of every chemical element a characteristic distance, chosen so that this relation agrees with experimental data.

It is possible to expand this model considerably by taking a semi-classical approach, in which momentum is quantized. This approach works very well for the hydrogen atom, which only has one electron. The magnitude of the angular momentum for a circular orbit is. The total energy of the atom is the sum of the kinetic and potential energies, that is.

Velocity can be eliminated from the kinetic energy term by setting the Coulomb attraction equal to the centripetal force, giving. Now the energy can be found in terms of Z, e, and r. Using the new value for the kinetic energy in the total energy equation above, it is found that:

Solving the angular momentum for v and substituting this into the expression for kinetic energy, we have. This establishes the dependence of the radius on n. *That is*: At its smallest value, n is equal to 1 and r is the Bohr radius a''_0. Now, the equation for the energy can be established in terms of the Bohr radius.

QUANTUM-MECHANICAL EXPLANATION

According to the more complete theory of quantum mechanics, the location of an electron is best described as a probability distribution. The energy can be calculated by integrating over this cloud. The cloud's underlying mathematical representation is the wavefunction which is built from Slater determinants consisting of molecular spin orbitals. These are related by Pauli's exclusion principle to the antisymmetrized products of the atomic or molecular orbitals. In general, calculating the nth ionization energy requires calculating the energies of and electron systems. Calculating these energies is not simple, but is a well-studied problem and is routinely done in computational chemistry.

At the lowest level of approximation, the ionization energy is provided by Koopmans' theorem.

VERTICAL IONIZATION ENERGY

The geometry of a molecular ion may be different from the neutral molecule. The measured ionization energy can refer to the vertical ionization energy, in which case the ion has the same geometry as the neutral molecule, or to the adiabatic ionization energy, in which case the ion has its lowest energy, relaxed geometry. This is illustrated in the figure. For a diatomic molecule the only geometry change possible is the bond length. The figure shows an ion with a slightly longer bond length than the neutral molecule. The harmonic potential energy surfaces are shown in green (neutral) and red (ion) with vibrational energy levels. The vertical ionization energy is always greater than the adiabatic ionization energy.

RYDBERG CONSTANT

The most accurately measured of the fundamental constants; it is a universal scaling factor for any spectroscopic transition and an important cornerstone in the determination of other constants. This constant was introduced empirically. J. Balmer's formula described the visible spectral lines of atomic hydrogen, while J. Rydberg's formula applied to the spectra of many elements. Their results may be summarized by Eq. (1),

$$\frac{1}{\lambda} = R\left(\frac{1}{n_1^2} - \frac{1}{n_2^2}\right) \qquad (1)$$

where λ is the wavelength of the spectral line and R is a constant. In Balmer's account of the visible hydrogen spectrum, n_1 was equal to 2, while n_2 took on the integer values 3, 4, 5, and so forth. In Rydberg's more general work, n1 and n2 differed slightly from integer values. A remarkable result of Rydberg's work was that the constant R was the same for all spectral series he studied, regardless of the element. This constant R has come to be known as the Rydberg constant. Applied to hydrogen, Niels Bohr's atomic model leads to Balmer's formula with a predicted value for the Rydberg constant given by Eq. (2),

$$R\infty = \frac{m_e e^4}{8h^3 \varepsilon_0^2 c} \qquad (2)$$

where me is the electron mass, e is the electron charge, h is Planck's constant, ε_0 is the permittivity of vacuum, and c is the speed of light. The equation expresses the Rydberg constant in SI units. To express it in cgs units, the right-hand side must be multiplied by (4&pgr;ε_0)^2. The subscript ∞ means that this is the Rydberg constant corresponding to an infinitely massive nucleus. E. Schrödinger's wave mechanics predicts the same energy levels as the simple Bohr model, but the relativistic quantum theory of P. A. M. Dirac introduces small corrections or fine-structure splittings. The modern theory of quantum electrodynamics predicts further corrections. Additional small hyperfine-structure corrections account for the interaction of the electron and nuclear magnetic moments. Fine structure (spectral lines), Hyperfine structure

The Rydberg constant is determined by measuring the wavelength or frequency of a spectral line of a hydrogenlike atom or ion. The highest resolution and accuracy has been achieved by the method of Doppler-free two-photon spectroscopy, which permits the observation of very sharp resonance transitions between long-living states. The 2002 adjustment of the fundamental constants, taking into account different measurements, adopted the value R∞ = 10,973,731.568,525 ± 0.000,073 m^{-1} for the Rydberg constant. The measurements provide an important cornerstone for fundamental tests of basic laws of physics. Atomic structure and spectra, Fundamental constants, Laser, Laser spectroscopy

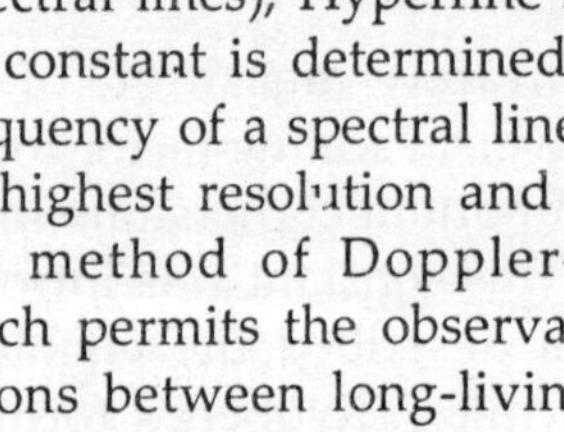

Illustration courtesy NIST:

- In a Rydberg atom, an electron (black dot) is far away from the atomic nucleus (red and grey core).
- Probability map for an electron in a Rydberg atom shows that it has virtually no probability of being near the nucleus in the centre.
- An optical frequency comb for producing ultraprecise colours of light can trigger quantum energy jumps useful for accurately measuring the Rydberg constant.

Researchers at the National Institute of Standards and Technology (NIST) and Max Planck Institute for Physics in Germany believe they can achieve a significant increase in the accuracy of one of the fundamental constants of nature by boosting an electron to an orbit as far as possible from the atomic nucleus that binds it. The experiment, outlined in a new paper,[2] would not only mean more accurate identifications of elements in everything from stars to environmental pollutants but also could put the modern theory of the atom to the most stringent tests yet.

The physicists' quarry is the Rydberg constant, the quantity that specifies the precise colour of light that is emitted when an electron jumps from one energy level to another in an atom. The current value of the Rydberg constant comes from comparing theory and experiment for 23 different kinds of energy jumps in hydrogen and deuterium atoms. Researchers have experimentally measured the frequencies of light emitted by these atomic transitions (energy jumps) to an accuracy of as high as 14 parts per quadrillion (one followed by 15 zeros), but the value of the Rydberg constant is known only to about 6.6 parts in a trillion–500 times less accurate. The main hurdle to a more accurate value comes from uncertainties in the size of the atom's nucleus, which can alter the electron's energy levels and therefore modify the frequency of light it emits. Another source of uncertainty comes from the fact that electrons sometimes emit and reabsorb short-lived"virtual photons," a process that also can slightly change the electron's energy level.

HYDROGEN AND RYDBERG CONSTANT

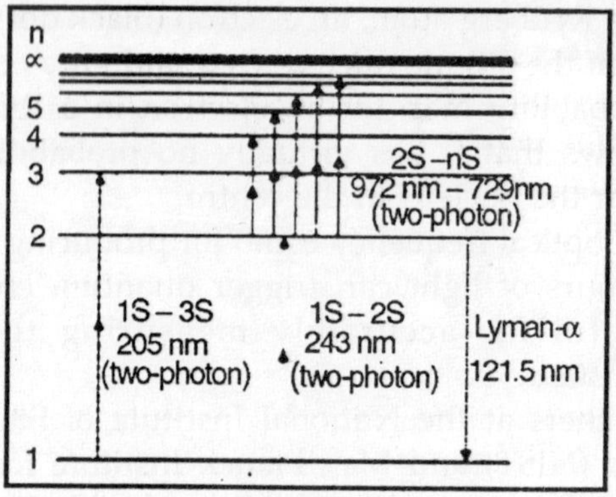

Energy Levels in Atomic Hydrogen

An experiment is underway at NPL to make accurate measurements of the frequency of a series of two-photon transitions in atomic hydrogen. This work will contribute towards an improved determination of the Rydberg constant.

Hydrogen is unique among optical frequency standards in providing a range of calculable reference frequencies, from radio frequencies into the vacuum ultraviolet region, which are linked by well-defined atomic theory and are provided by the same atomic system. The scaling factor for these transition frequencies is the Rydberg constant, which is currently known to 6.6 parts in 10^{12} [1]. Although the 1S-2S transition frequency in hydrogen has been measured to 1.4 parts in 10^{14} [2], this is not sufficient to determine the Rydberg constant with higher accuracy because of quantum electrodynamic and nuclear size corrections to the energy levels. To separate out these effects additional transition frequencies must be measured.

In our experiment we are studying the 2S-nS,D transitions by Doppler-free two-photon laser spectroscopy on a metastable hydrogen atomic beam[3]. In the atomic beam, metastable atoms are produced by electron impact excitation and detected by electric field quenching of the metastable state. To induce the 2S-nS,D transitions we use a Ti:sapphire laser which is stabilized to a high-finesse ultralow-expansion cavity and which is collinear with the metastable atomic beam. A build-up cavity constructed around the atomic beam produces the counter-propagating beams required for the Doppler-free two-photon excitation and increases the excitation rate. The transitions are detected by monitoring the reduction in the number of metastable atoms reaching the far end of the beam as the laser frequency is swept through the resonance.

We have recently observed the first laser-induced signals from our apparatus and made a preliminary measurement of the $2S_{1/2}$-$8D_{5/2}$ transition frequency using a hydrogen-maser-referenced femtosecond optical frequency comb. Detailed studies of systematic frequency shifts are in progress. The femtosecond comb provides the capability to study all optical transitions with $n > 4$ under nominally identical conditions. This enables us to carry out more stringent tests of a number of systematic effects than in previous experiments.

The limiting systematic effects in this first experiment are expected to be ac Stark shifts and saturation effects associated with the high laser power required to induce the two-photon 2S-nS,D transitions. To reduce these effects, we propose in a later stage of the experiment to produce the metastable 2S state by two-photon optical excitation from the ground state using a stabilised 243 nm laser source. Optical excitation, in contrast to electron impact excitation, imparts no additional momentum to the atoms and so should generate an intense, well-collimated source of metastable atoms, which can be cooled to increase further the atomic density in the beam. This in turn would enable the 2S-nS,D transitions to be excited using lower laser power, reducing the light shifts. A separate apparatus is being constructed to test the optical excitation efficiency, while electron impact excitation continues to be used for initial experiments in the primary apparatus.

Rydberg's Constant for Hydrogen was determined to be about 11,214,870m^{-1}, resulting in only about a 2.25% discrepancy with the presently accepted value of R_H1 using a spectrometer and a diffraction grating.

Long before Bohr developed his orbital model of the atom and derived the general equation:

$$1/\lambda = R_H(1/(n_f2)-1/(n_i2))$$

relating the wavelength of the light emitted by an excited atom to its orbital transition, Balmer had already determined that a similar equation,

$$1/\lambda = k((1/4)-1/(m^2)),$$

where m is an integer and k is a constant, accurately described the spectroscopic lines when visible light was emitted. Balmer was able to come up with his formula empirically due to the abundance of data available for visible emissions.

The scattering from a set of Bragg planes establishes the Bragg condition for constructive interference, referred to in the context of this experiment as the grating equation, that

$$n\lambda = 2d\sin(\theta),$$

where n is the order of the observed diffraction, λ is the wavelength of the incident light, d is the spacing of the diffraction grating, and θ is the angle through which the light is diffracted.

Experiment

The ease of measurement of diffractions in the visible spectrum due to the ability of the human eye to observe them was utilized in this experiment. The visible spectra emitted by excited sodium and hydrogen atoms were observed using a spectrometer in order to determine the diffraction angles of each atom's various emissions in both the first and second order through a diffraction grating.

For the sodium, the presently accepted values for the wavelengths of the sodium emissions2 were taken as known quantities. This enabled the calculation of the spacing in the diffraction grating using the grating equation. The mean of the calculated values for the spacing of the diffraction grating was taken to be the spacing of the diffraction grating for the remainder of the experiment. For the hydrogen, the wavelengths of the various spectral lines were calculated for each measurement using the grating equation. Four spectral lines were observed. The diffraction angles for the red, turquoise, and light purple spectral lines were observed and recorded in both the first and second order as they bent both to the left and right side of the spectrometer. The diffraction angles for the dark purple line, however, could only be recorded in the first order as it bent to the left and right side of the spectrometer.

Using Bohr's orbital transition model for atomic emissions each observed value of λ could reasonably be paired with its orbital transition. The orbital transitions for each spectral line were always to the second orbital because if they were not they would not have been observable in the visible spectrum. The orbital from which this transition was made was different for each spectral line, increasing incrementally from 3 to 6 as the colour changed from red to dark purple. This was justified because a more energetic emission requires a larger orbital transition, and light has more energy at shorter wavelengths. This pairing led to14 data points of the form $((1/(n_f2)-1/(n_i2)), 1/\lambda)$. Comparing this plot to equation, it is easy to see that, since equation (1) is linear with regards to the quantities being plotted, a linear regression on these data points should yield a line with slope R_H. The value of R_H thus obtained was taken to be the experimentally calculated value of R_H.

Analysis and Discussion

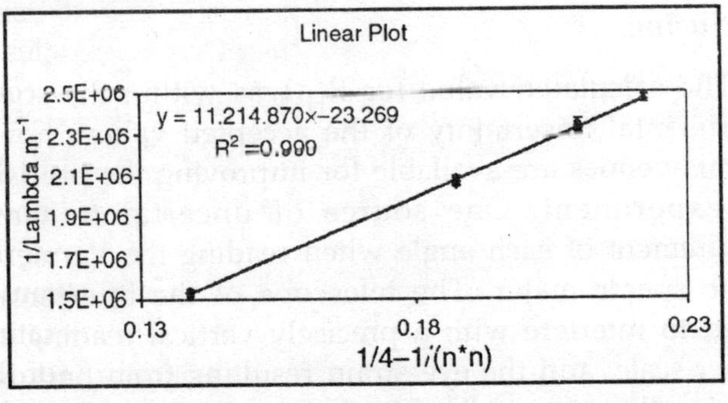

Fig. Linear Plot of the Hydrogen Data According to Balmer's Formula (1). The Slope of the Line Obtained Via Linear Regression is the Experimentally Obtained value of R_H.

Using a least-squares linear regression on the 14 data points, the linear fit shown in was obtained. The slope of the best-fit line is 11,214,870m^{-1}, compared with the accepted value1 for RH of 10,967,760m^{-1}, an absolute discrepancy of 247,110m^{-1}, but a per centage discrepancy of only 2.25%. The linear fit was also fantastic, with r^2 =.9990 and correlation coefficient r =.9995.

The linear fit, however, has a nonzero additive constant of -23,269m^{-1}. Fortunately, in analogy with the absolute discrepancy, this is small relative to the enormity of the slope, being only about 0.2% as large.

The uncertainty in the measurement of diffraction angles was only.1°; however, since the measurement of the wavelength of the light in each case was dependent upon both the diffraction grating spacing and the diffraction angle, and the measurement of the diffraction grating spacing was also dependent upon the angle of diffraction, the uncertainty in the wavelength of the light in each case was doubly dependent upon the uncertainty in the angle. Such error propagation analysis enabled error bars to be included on the linear plot. This allowed a maximum and minimum value of R_H to be estimated by calculating a maximum and minimum reasonable slope of the linear fit. Taking points (5/36, 1500000) and (2/9, 2500000), and (5/36, 1600000) and (2/9, 2400000) leads to R_{Hmax} = 12,000,000m^{-1} and R_{Hmin} = 9,600,000m^{-1}.

Conclusion

The calculated value for R_H was within the specified experimental uncertainty of the accepted value[1]; however, several avenues are available for improving the precision of this experiment. One source of uncertainty was the measurement of each angle when reading the Vernier scale on the spectrometer. The telescope of the interferometer tended to interfere with a precisely vertical reading of the Vernier scale, and the eye strain resulting from finding and centreing the spectral lines in the eyepiece compounded the difficulty.

A finely calibrated digital display would alleviate these

problems. A method for focusing the spectrometer automatically and with less uncertainty would also make the determination of the centre of each of the spectral lines in the eyepiece more precise.

Conducting the experiment in a pitch black dark room would also ease the process of finding and centreing spectral lines, possibly enabling the detection of the second order dark purple line, though data collection would be impeded in such a situation.

SOLAR SPECTRUM AND FRAUNHOFER LINES

SOLAR SPECTRUM

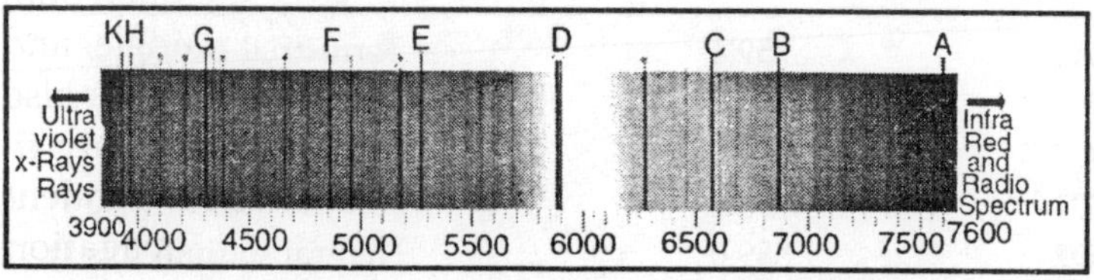

BACKGROUND AND THEORY

The brightest star in our sky is the Sun. Absorption lines in the solar spectrum were first noticed by an English astronomer in 1802, but it was a German physicist, Joseph von Fraunhofer, who first measured and cataloged over 600 of them about 10 years later.

These lines are now known collectively as the"Fraunhofer lines." In the 1800's, scientists did not know that these lines were chemical in origin. Thus, the letters used by Fraunhofer to identify the lines have no relation to chemical symbols nor to the symbols used to designate the spectral types of stars. Today's astronomers use some of the designations simply for convenience and ease in identifying lines.

Now we know that each absorption line is caused by a transition of an electron between energy levels in an atom. Each element has a distinct pattern of absorption lines. Once the pattern of the lines of a particular element have been observed in the laboratory, it is possible to determine whether those elements exist elsewhere in the universe simply by pattern matching the absorption lines.

The strongest Fraunhofer lines of the Sun can easily be seen with even the most primitive spectroscope. By viewing a bright sky or the full Moon with a spectroscope or diffraction grating, you should be able to see at least a few absorption lines. In this exercise, we work with the solar spectrum between approximately 390 and 660 nm (3900 - 6600 Angstroms) and identify some of the strongest Fraunhofer lines.

Procedure

Table. --"Known" Lines

Designation	Wavelength (nm)	Origin
A	759.4	Terrestrial oxygen
B	686.7	Terrestrial oxygen
C	656.3	Hydrogen (Hα)
D1	589.6	Neutral sodium (Na I)
D2	589.0	Neutral sodium (Na I)
E	527.0	Neutral iron (Fe I)
F	486.1	Hydrogen (Hβ)
H	396.8	Ionized calcium (Ca II)
K	393.4	Ionized calcium (Ca II)

- On the solar spectrum, measure the distance (in pixels) between two widely spaced,"known" lines.
- Find the distance between these lines in nanometers using.
- Divide the distance in nanometers (step 2) by the distance in pixels to get the number of nanometers per pixel.
- Average the results of the four measurements to get the scaling factor.

Calculate the wavelengths of the"unknown" lines completing Table:

- Pick one of the lines to serve as your reference; for example, the K line of Ca II at 393.4 nm. Measure the distance from this line to each of the numbered lines (1-13) in pixels.

- Use the scaling factor and convert the distances in pixels to distances in nm.
- Add the distance in nm to the wavelength of your reference line to get the wavelength of the"unknown" lines.7
- Compare these wavelengths to the list of lines in and identify the"unknown" lines. Your values may not exactly match those given in the table. Why?

Note: If you find that some of your calculated wavelengths do not seem to match any of those in the table, find the closest match and the corresponding element."Flag" these wavelengths in.

PROMINENT FRAUNHOFER LINES

Fraunhofer lines are dark absorption lines in the solar spectrum that can be seen when sunlight is passed through a prism to separate it into the colours of the rainbow. They occur because cooler gas, which is higher in the Sun's atmosphere, absorbs some colours of the light emitted by hotter gas lower in the Sun's atmosphere. Sir Isaac Newton (1642-1727) discovered that if white light is passed through a prism, it separates into a rainbow, which is called a spectrum.

While studying the spectrum that sunlight made, Joseph Fraunhofer (1787-1826) discovered some dark lines scattered among the colours. These dark lines were segments of colours missing from the complete spectrum. Fraunhofer counted 574 of these lines, which we now call Fraunhofer lines. Today, using much more sophisticated techniques, astronomers have discovered tens of thousands of Fraunhofer lines. Why doesn't the Sun emit these missing colours? Or, if the Sun does emit these colours, what happens to the light before it reaches Earth? The answer lies at the surface of the Sun.

When we look at a picture of the Sun, the surface that we see is called the photosphere. The photosphere is a region, several hundred kilometres thick, in which the Sun changes from opaque to transparent. It is not actually the outermost surface: the Sun extends for thousands of kilometres beyond the photosphere, but it is not usually visible from Earth. The photosphere is interesting because within this thin layer of the Sun (thin compared to the whole Sun, of course) sunlight is created, and some of the colours are lost almost immediately.

The lower region of the photosphere has a temperature of about 10,000° F (about 5,500°C) and glows white-hot. Any object that glows due to a high temperature gives off a complete spectrum, that is, it has all the colours of the rainbow. As this light proceeds upwards in the Sun into a higher region of the photosphere, the temperature drops several thousand degrees. Although most of the light passes right through, some of the light is absorbed by the cooler gas. Only certain colours are removed because the chemical elements in the photosphere can only absorb certain wavelengths of light, and different wavelengths correspond to different colours. For example, sodium absorbs some yellow light at a wavelength of about 5.89×10^{-7}m. These absorbed colours cause the Fraunhofer lines. By measuring precisely the wavelengths of the missing colours, that is, the Fraunhofer lines, and how much light is actually absorbed, astronomers have learned much about the temperature inside the Sun and its chemical composition.

We can also learn about other stars in the sky by looking at the absorption lines in their spectra. By studying the similarities and differences that they have with the Fraunhofer lines, we can learn a lot about the similarities and differences that other stars have with our Sun.

Absorption lines in the spectrum of the Sun, or of another star, first studied and named by Joseph von Fraunhofer in 1814. Altogether, Fraunhofer found some 700 lines in the solar spectrum. The nine most prominent he labeled with capital letters A to K, starting at the red end. The A and B bands are now known to be caused by absorption in Earth's atmosphere, while the rest are due to absorption in the Sun's photosphere. C and F are now better known as H-alpha and H-beta; the D lines are of sodium, the H and K lines of calcium, and the G band by neutral iron and the CH molecule. All these features occur generally in stars of spectral types F, G, and K.

Lines	Due To	Wavelengths
A - (band)	O_2	7594 - 7621
B - (band)	O_2	6867 - 6884
C	H	6563
a - (band)	O_2	6276 - 6287
D - 1, 2	Na	5896 & 5890
E	Fe	5270
b - 1, 2	Mg	5184 & 5173
c	Fe	4958

Lines	Due To	Wavelengths
F	H	4861
d	Fe	4668
e	Fe	4384
f	H	4340
G	Fe & Ca	4308
g	Ca	4227
h	H	4102
H	Ca	3968
K	Ca	3934

FLUORESCENCE AND PHOSPHORESCENCE

These phenomena are closely related to electronic absorption spectra and can be used as a tool for analysis and structure determination. Both involve the absorption of radiation via an electronic transition, a loss of energy through either vibrational energy decay or nonradiative processes, and the subsequent emission of radiation of a lower frequency than that absorbed.

Electrons possess intrinsic magnetic moments that are related to their spin angular momenta. The spin quantum number is $s = {}^1/_2$, so in the presence of a magnetic field an electron can have one of two orientations corresponding to magnetic spin quantum number $ms = \pm{}^1/_2$. The Pauli exclusion principle requires that no two electrons in an atom have the same identical set of quantum numbers; hence when two electrons reside in a single AO or MO they must have differentms values (i.e., they are antiparallel, or spin paired). This results in a cancellation of their magnetic moments, producing a so-called singlet state. Nearly all molecules that contain an even number of electrons have singlet ground states and have no net magnetic moment (such species are called diamagnetic). When an electron absorbs energy and is excited to a higher energy level, there exists the possibility of:

- Retaining its antiparallel configuration relative to the other electron in the orbital from which it was promoted so that the molecule retains its singlet characteristic,
- Changing to a configuration in which its magnetic moment is parallel to that of its original paired electron.

In the latter case, the molecule will possess a net magnetic moment (becoming paramagnetic) and is said to be in a triplet state. For each excited electronic state, either electron spin configuration is possible so that there will be two sets of energy levels (see Figure).

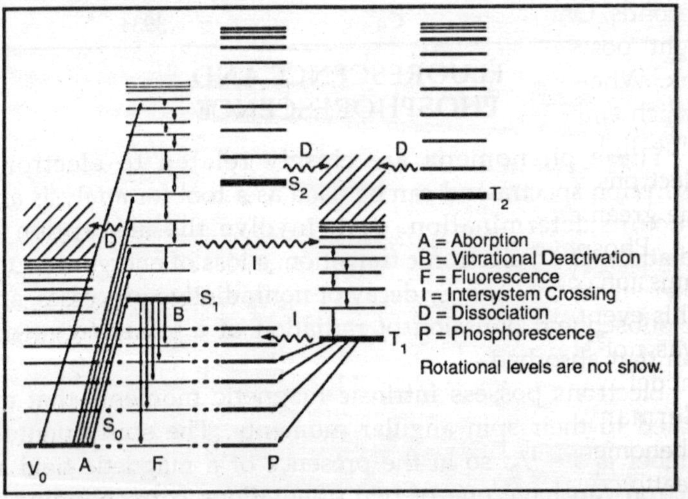

Fig. Energy-level Diagram and Possible Transitions for a Polyatomic Molecule Having a Singlet, S_0, Ground State and Both Singlet, S_1 and S_2, and Triplet, T_1 and T_2, Excited States.

A = absorption,
B = vibrational deactivation,
F = fluorescence,
I = intersystem crossing,
D = dissociation,
P = phosphorescence.
Rotational levels are not shown.

Fluorescence and phosphorescence are both luminescencephenomena that occur due to the de-excitation of electrons. When a source of energy such as thermal or radiant energies is applied to a fluorescent material, high energy particles or photons collide with the electrons in the material, causing them to become excited and jump to a higher energy level. The electrons in the excited state have the same spin as they did in the ground state. The electrons in this excited state are not stable and almost spontaneously become de-excited and move back down to the lower energy levels, releasing energy in the form of light.

When the source of energy is removed, the light emission ceases. A fluorescent state can exist for only 10^{-5} to 10^{-8} seconds. One familiar example of this phenomenon is a"black light" poster."Black light" posters are painted with fluorescent ink. When the ink is exposed to"black light" (a light bulb which emits mostly blue and ultraviolet light), it absorbs the blue light, which excites the electrons in the ink. As the electrons move back to their ground state, they emit light in the green and yellow wavelengths.

Phosphorescence is similar to fluorescenceexcept the light emission continues even after the source of energy is removed. This event was first observed in the seventeenth century but was not studied until the nineteenth century. The current model was proposed by Philipp Lenard of Kiel University, Germany. He had been interested inluminescent t phenomenon since he was a boy in Hungary and would apply heatto crystals of fluorine to make them glow.

In 1888, working at the University of Heidelberg, he began work with cathode rays and found that they were not equivalent to fluorescent light, as was the current thinking.

He worked for many years on thephotoelectric effect, trying to find an explanation for the phenomenon. In 1902, he published his theory of electronexcitation and luminescence. His theory could not be explained, however, until 1905 when Albert Einstein developed the theory of photons. Lenard won the 1905 Nobel prize in physics for his work in cathode rays.

A phosphorescent material emits light after it absorbs energy, for instance, radiant energy. The absorbed energy excites electrons in the phosphorescent material and causes them to be caught inpotential energy troughs. While excited, the electrons experience a change in spin called intersystem crossing.

This change in spin places the electrons in a metastable position from which they cannot move back into the ground state. They are freed from these troughs by thermal energy within the crystal structure of the material, which causes them to be raised to a higher, less stable energy level where they can eventually fall back to lower energy levels. As they do so, they emit light. Because of the slow decay due to the intersystem crossing, the light continues after the exciting source is removed.

Phosphorescence is temperature dependent since thermal energyis required for the electrons to become de-excited.Phosphorescent states can exist from 10^{-4} seconds to several hours. Phosphorescent materials are used in"glow-in-the-dark" toys that will emit light after exposure to some form of radiant energy.

The hands on certain watches and alarm clocks are coated with a phosphorescent material which will emit light for many hours after a light source is removed. The most common form of phosphorous exhibits this phenomenon, hence the term"phosphorescence." Some fluorescent materials, under the appropriate conditions, will also phosphoresce.

Fluorescence has been utilized for many purposes, the most obvious being fluorescent lamps. A fluorescent lamp is an evacuated glass tube that contains mercury vapour and is coated with an apatite analog ($Sr_5(PO_4)_3F$), small amounts of rare earth elements, and antimony, which is used as an indicator. When a voltage is applied across the tube (approximately 600 Volts), the mercury ionizes and conducts. An electron is stripped from a mercury atom.

When the stripped electron collides with an electron in another mercury atom, it causes the second electron to become excited and jump to a higher energy level. When that electron returns to its ground state, it releases energy as ultraviolet light, which is invisible.

The coating on the inside of the tube is referred to as phosphors. Phosphors absorb ultraviolet energy, causing the electrons in the material to become excited. When these electrons return to their ground state, they release light in the visible region.

Phosphors are also used in laundry detergents to make white clothes appear whiter. When the ultraviolet radiation from sunlight is absorbed the phosphors in theclothes glow. Sometimes a mixture of krypton and argon are used instead of the mercury.

There are many naturally occurring fluorescent materials. Many minerals, for example, will emit light when activated by ultraviolet light, x rays, or electron beams. Electrons in the mineral absorb the energy from the activator and become excited. As they return to their ground state, they emit light. The light emitted after activation is sometimes a very different colour from the mineral itself.

Many mineral collectors use fluorescence to identify minerals. A special fluorescent lamp that emits ultraviolet light while filtering out white light is required to view the fluorescence in many minerals. Some minerals are considered fluorescent even though not all samples of the mineral will fluoresce.

This is often the result of the different samples of the mineral being obtained from different locations. Many times impurities in the sample will inhibit, or change the colour of, the mineral's fluorescence.

Sometimes the colour of a mineral's fluorescence can be used to determine its place of origin. For example, calcite fluoresces yellow, white, green, red, or just about any of the colours of the visible spectrum; but if it is from Terlingua, Texas, it is always bluish-white and if it is from Franklin, New Jersey, it is always red.

Two kinds of ultraviolet light can be used to detect fluorescence in minerals, long wave and short wave. Some minerals fluoresce the same under both wavelengths, some fluoresce different colours under each wavelength, and some only fluoresce under one of the two types of ultraviolet light. Some minerals which fluoresce under long wave ultraviolet light include agate, calcite, fluorite, and halite, which produce a white light; amber, which produces an orange light; and quartz and opal, which produce green light.

Minerals which fluoresce under short wave ultraviolet light include gypsum, which produces white light; ruby, which produces red light; talc, which produces yellow light; and diamond, which produces green light. Fluorescent diamonds were once considered more precious than non-fluorescent stones, although recently the fluorescent property is considered an impurity.

Fluorescence and phosphorescence are two similar luminescentphenomena. Fluorescence is the emission of light from a material while being exposed to an energy source, while phosphorescenceis the emission of light from a material during, as well as after, exposures to an energy source. The difference between the two is the result of the spin of the excited electrons.

There are many familiar examples of fluorescence and phosphorescence in our daily lives. Fluorescent materials can be synthetic, such as phosphors, or occur in nature, for example, fluorescent minerals.

FLUORESCENCE OF QUININE SULFATE

Quinine, in the form of its sulfate salt, is a flavour component of tonic water. The bitter taste of anti-malarial quinine tonic led British colonials in India to drink it mixed with gin, thus creating the gin and tonic cocktail. Today, many soft drinks contain a small amount of the substance as a flavouring agent.

Upon excitation by UV light, tonic water glows with blue light due to the fluorescence of quinine. Figure shows the well-known Jablonski diagram] illustrating the related phenomena of absorption, fluorescence, and phosphorescence.

Fluorescence occurs when the first excited singlet state (S_1) returns to the ground singlet state (S_0) by emitting a photon. The emission spectrum has a maximum at about 450 nm, accounting for the bluish colour. A very crude upper-limit estimate of the fluorescence lifetime, as defined by the time it takes to decrease the fluorescence intensity to 1/e of its original value, can be obtained by quickly blocking the excitation light source and noting the instant disappearance of fluorescence, confirming that the lifetime must be much less than a second. In reality, the fluorescence lifetime of quinine sulfate is about 19 ns.

FLUORESCENCE OF CHLOROPHYLL

A clear green chlorophyll (Chl) solution can be made by grinding spinach (either from fresh or frozen samples) in a mortar and pestle, adding the broken leaves to a beaker of acetone, and then filtering the solution through cheesecloth or coarse filter paper to remove the solid plant materials. Chlorophyll a and chlorophyll b molecules, both present in plants, impart the green colour to the solution. When irradiated with UV light, these molecules fluoresce at about 650 nm to give a beautiful red solution. As with quinine sulfate, the mean fluorescence lifetime for both compounds is very short - about 5 ns.

When present in plant cells, instead of fluorescing, an electronically excited chlorophyll molecule (Chl*) transfers an electron to a nearby acceptor molecule (A) as follows:

Chl* + A → Chl+ + A- (1)

A- → A + e- (2)

The electron is then passed along an electron-transport chain containing a quinone and other molecules to reach the reaction centre. Eventually, the excess energy originally captured by the chlorophyll molecules is used to synthesize carbohydrates using atmospheric CO_2 as a carbon source [4,

5].An interesting extension of this demonstration with chlorophyll is to introduce fluorescence quenching. In solution, we can demonstrate the quenching of chlorophyll fluorescence by subsequently adding 2,5-dimethyl-p-benzoquinone (BQ), which is an effective electron acceptor. In the absence of the electron-transport chain present in plants, the ion-pair formation process shown in Equation (1) undergoes rapid spin-allowed electron back transfer to yield ground state reactants:

Chl* + BQ → Chl+ + BQ- (3)

Chl+ + BQ- → Chl + BQ (4)

This process is responsible for quenching the fluorescence.

PHOSPHORESCENCE OF NAPHTHALENE

Like quinine and chlorophyll, the aromatic hydrocarbon naphthalene ($C_{10}H_8$), which bears a superficial resemblance to the porphyrin ring of chlorophyll, can undergo fluorescence although most of the emitted light is invisible because it is in the UV region. As Figure shows, the first excited singlet state (S_1) of naphthalene can also decay by an alternate radiationless transition to a triplet state (T_1) of lower energy (a phenomenon known as intersystem crossing). During this process, some energy is lost as heat to the surrounding molecules. Once a molecule is in the triplet state, it can lose its excess energy and return to the ground state (S_0) either by spontaneously emitting a photon (this process is called phosphorescence), or undergoing another radiationless transition. Because the triplet to singlet transition is spin-forbidden, the lifetime of a triplet state is usually found to persist for seconds, which is much longer than that for fluorescence, as the previous examples demonstrate. To avoid deactivation of the triplet state by collision with other molecules, phosphorescence spectra are usually recorded with samples prepared in a frozen medium.

In this demonstration, we use naphthalene to show the effect of deuteration (that is, $C_{10}D_8$ versus $C_{10}H_8$) on the lifetime of

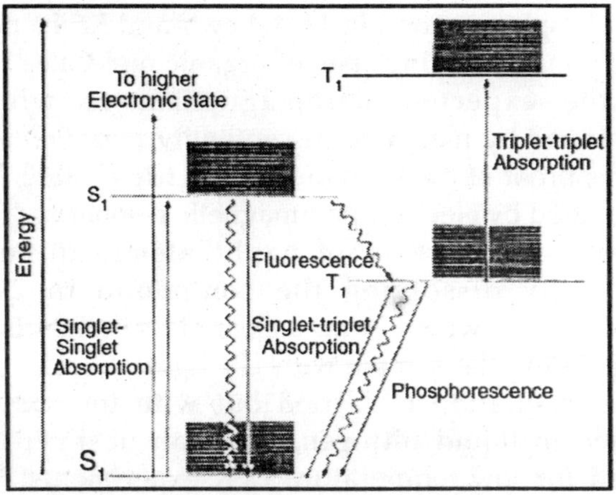

Fig. Jablonski Diagram Showing Absorption, Fluorescence, and Phosphorescence. The Wavy Lines Indicate Radiationless Transitions; the Closely Spaced Lines Represent the Vibrational Levels.

Fig. Left: Fluorescence of Tonic Water. Right: An Acetone Solution of Chlorophyll under Direct UV Irradiation.

its phosphorescence. In 1944, Lewis and Kasha postulated that the phosphorescing state of organic molecules must be a triplet,the expected paramagnetism of which was demonstrated by magnetic susceptibility measurements. The definitive proof of the formation of the triplet state, however, was provided by electron paramagnetic resonance (EPR). The green phosphorescence of naphthalene can be readily observed by dissolving the compound in 2-methyl-tetrahydrofuran, which forms a clear glass in the frozen state, and irradiating the sample with UV light.

The irradiation is carried out with the sample tube immersed in liquid nitrogen, using an unsilvered Dewar designed for low-temperature EPR experiments. With the $C_{10}H_8$ sample, the decay of the resulting phosphorescence, as evidenced by the diminishing green light, persists for a few seconds after the excitation light source is blocked. A dramatic increase in the phosphorescence lifetime is observed when $C_{10}D_8$ is used as the sample. In this case, the emitted green light can be observed even after 30 seconds! (The literature values for the phosphorescence lifetimes are 2.7 s for $C_{10}H_8$ and 22 s for $C_{10}D_8$, respectively.)

The reason for the dramatic increase in the phosphorescence lifetime from $C_{10}H_8$ to $C_{10}D_8$ is due to the fact that the relative values of the radiative and radiationless transition probabilities vary considerably between these two molecules. It is found that the complete substitution of deuterium for hydrogen in aromatic hydrocarbons increases both the observed lifetime of the phosphorescence and its quantum yield.

However, there is relatively little difference in the energy of these transitions or the nature of the respective excited triplet states, as shown by the similar appearance of the phosphorescence and EPR signals for $C_{10}H_8$ and $C_{10}D_8$. Hence, it has been concluded that deuteration has altered only the relative rate of the radiationless transition. Most theories predict that only high-energy vibrations will be effective in promoting radiationless transitions between the lowest triplet state and the ground state.

In aromatic hydrocarbons, the highest-energy vibrations are the C-H stretching motions, which occur at frequencies of about 3,000 cm^{-1}.

Complete deuteration decreases the frequency of these vibrations by roughly the square root of the ratio of the proton mass to the deuteron mass or to a frequency of about 2,200 cm^{-1}.

This lower vibrational frequency is believed to be the major cause of the decrease of the rate constant for the radiationless process upon deuteration, and hence the observed longer lifetime of the triplet state.

The above demonstrations are fairly easy to carry out in either a small classroom or a large lecture hall, where a video camera and projection screen may be used to show the light emissions. Although the underlying principles require some knowledge of quantum mechanics, we believe that parts of the demonstrations, namely the fluorescence of quinine and chlorophyll and the quenching of chlorophyll fluorescence, are also suitable for the general chemistry audience.

EXPERIMENTAL

UV irradiation was provided by a short wavelength (254 nm) hand-held UV lamp (model UVG-1), or by a larger UV lamp ("BLAK-RAY" Longwave Ultraviolet Lamp, model B-100A), both from UVP, Inc.

Solvents and chemicals [CAS number] (naphthalene, perdeutero-naphthalene [1146-65-2], and 2-methyltetrahydrofuran were purchased from Sigma-Aldrich and were used without additional purification. 2,5-Dimethyl-p-benzoquinone was obtained from Fisher Scientific/Acros-Organics. Commercially available tonic water was used directly. Frozen spinach was thawed to room temperature before extraction of chlorophyll.

A freshly prepared acetone extract of chlorophyll was prepared by crushing ca. 150 g of spinach (after thawing frozen spinach, it should be drained through a metal screen; fresh spinach may also be used) in a medium mortar and pestle. The crushed spinach was transferred to a 250 mL beaker and extracted by swirling with 200 mL of acetone. The acetone extract was then filtered through coarse filter paper (cheesecloth may be substituted) into a 250 mL beaker. A small aliquot (2-3 mL) of this solution is diluted with acetone and the intensity of the resulting red fluorescence upon direct irradiation with the UV lamp is observed.

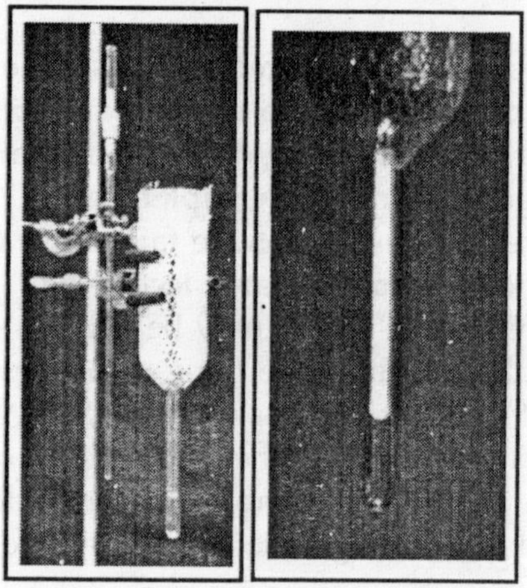

Fig. *Left*: EPR tube and EPR Dewar. *Right*: Phosphorescence of a frozen solution of $C_{10}D_8$ in 2-methyltetrahydrofuran after brief UV irradiation.

Progressive serial dilution should be carried out until a sample with suitably intense fluorescence is obtained. Use of a Petri dish to display this fluorescence is convenient; the tonic water fluorescence can be easily displayed by direct irradiation of the bottle.

A small sample of 2,5-dimethyl-p-benzoquinone (ca. 50 mg) can be gradually added to the acetone solution of chlorophyll under irradiation to demonstrate quenching. Darkening the room makes visualization of the fluorescence easier to observe. Chlorophyll/acetone solutions may be stored in the refrigerator and reused.

A solution of naphthalene, or perdeutero-naphthalene, was prepared by dissolution of 20 mg of the compound in 1 mL of 2-methyltetrahydrofuran in a small test tube. Degassing of solutions was not necessary. The solution was transferred to the EPR tube with a Pasteur pipet and the J. Young valve was sealed. The EPR Dewar was filled with liquid nitrogen and the EPR tube was slowly placed in the Dewar. After ca. 3 minutes, the sample was frozen solid. The sample was then briefly (ca. 15 sec) irradiated in a darkened room with the UV lamp before removing the lamp to allow the resulting phosphorescence to be observed. The sample can be used repeatedly with no noticeable deterioration.

Chapter 3

X-Rays and Productions

PHYSICS OF X-RAY

When fast-moving electrons slam into a metal object, x-rays are produced. The kinetic energy of the electron is transformed into electromagnetic energy. The function of the x-ray machine is to provide a sufficient intensity of electron flow from the cathode to anode in a controlled manner. The three principal segments of an x-ray machine - ca control panel, a high-voltage power supply, and the x-ray tube are all designed to provide a large number of electrons focused to a small spot in such a manner that when the electrons arrive at the target, they have acquired high kinetic energy.

Kinetic energy is the energy of motion. Stationary objects have no kinetic energy; objects in motion have kinetic energy proportional to their mass and the square of their velocity.

The equation used to calculate kinetic energy is:

$$KE = 1/2\ mv^2$$

where m is the mass in kilograms, v is the velocity in metres per second, and KE is the kinetic energy in joules. In determining the magnitude of the kinetic energy of a projectile, the velocity is more important than the mass.

In a x-ray tube, the projectile is the electron. As its kinetic energy is increased, both the intensity (number of x-rays) and the energy (their ability to penetrate) of the created x-rays are increased.

The x-ray machine is a remarkable instrument. It conveys to the target an enormous number of electrons at a precisely controlled kinetic energy. At 100 mA, for example, 6×10^{17}

electrons travel from the cathode to the anode of the x-ray tube every second.

The distance between the filament and the target is only about 1 to 3 cm. Imagine the intensity of the accelerating force required to raise the velocity of the electrons from zero to half the speed of light in so short a distance.

The electrons traveling from the cathode to anode in a vacuum tube comprise the x-ray current and are sometimes called projectile electrons.

When these projectile electrons impinge on the heavy metal atoms of the target, they interact with these atoms and transfer their kinetic energy to the target. These interactions occur within a very small depth of penetration into the target. As they occur, the projectile electrons slow down and finally come nearly to rest, at which time they can be conducted through the x-ray anode assembly and out into the associated electronic circuitry.

The projectile electron interacts with either the orbital electrons or the nuclei of target atoms. The interactions result in the conversion of kinetic energy into thermal energy and electromagnetic energy in the form of x-rays.

By far, most of the kinetic energy of projectile electrons is converted into heat. The projectile electrons interact with the outer-shell electrons of the target atoms but do not transfer sufficient energy to these outer-shell electrons to ionize them. Rather, the outer-shell electrons are simply raised to an excited, or higher, energy level. The outer-shell electrons immediately drop back to their normal energy state with the emission of infrared radiation. The constant excitation and restabilization of outer-shell electrons is responsible for the heat generated in the anodes of x-ray tubes.

Generally, more than 99% of the kinetic energy of projectile electrons is converted to thermal energy, leaving less than 1% available for the production of x-radiation. One must conclude, therefore, that, sophisticated as it is, the x-ray machine is a very inefficient apparatus.

The production of heat in the anode increases directly with increasing tube current. Doubling the tube current doubles the quantity of heat produced. Heat production also varies almost directly with varying kVp.

The efficiency of x-ray production is independent of the tube current. Regardless of what mA is selected, the efficiency of x-ray production remains constant. The efficiency of x-ray production increases with increasing projectile-electron endery. At 60 kVp, only 0.5% of the electron kinetic energy is converted to x-rays; at 120 MeV, it is 70%.

CHARACTERISTIC RADIATION

If the projectile electron interacts with an inner-shell electron of the target atom rather than an outer-shell electron, characteristic x-radiation can be produced. Characteristic x-radiation results when the interaction is sufficiently violent to ionize the target atom by total removal of the inner-shell electron. Excitation of an inner-shell electron does not produce characteristic x-radiation.

When the projectile electron ionizes a target atom by removal of a K-shell electron, a temporary electron hole is produced in the K shell. This is a highly unnatural state for the target atom and is corrected by an outer-shell electron falling into the hole in the K shell. The transition of an orbital electron from an outer shell to an inner shell is accompanied by the emission of an x-ray photon. the x-ray has energy equal to the difference in the binding energies of the orbital electrons involved.

EXAMPLE

A K-shell electron is removed from a tungsten atom and is replaced by an l_shell electron. What is the energy of the characteristic x-ray that is emitted?

Answer: For tungsten, K electrons have binding energies of 69.5 keV, and L electrons are bound by 12.1 keV. Therefore, the characteristic x-ray emitted has energy of:

$$69.5 - 12.1 = 57.4 \text{ keV}$$

In summary, characteristic x-rays are produced by transitions of orbital electrons from outer to inner shells. Since the electron binding energy for every element is different, the characteristic x-rays produced in the various elements are also different. This type of x-radiation is called characteristic radiation because it is characteristic of the target element. The effective energy characteristic x-rays increases with increasing atomic number of the target element.

DISCRETE X-RAY SPECTRUM

We saw earlier that characteristic x-rays have precisely fixed, or discrete, energies and that these energies are characteristic of the differences between electron binding energies of a particular element. A characteristic x-ray from tungsten, for example, can have one of fifteen energies and no others.

BREMSSTRAHLUNG RADIATION

The production of heat and characteristic x-rays involves interactions between the projectile electrons and the electrons of target atoms. A third type of interaction in which the projectile electron can lose its kinetic energy is an interaction with the nucleus of a target atom. In this type of interaction, the kinetic energy of the projectile electron is converted into electromagnetic energy.

A projectile electron that completely avoids the orbital electrons on passing through an atom of the target may come sufficiently close to the nucleus of the atom to come under its influence. Since the electron is negatively charged and the nucleus is positively charged, there is an electrostatic force of attraction between them. As the projectile electron approaches the nucleus, it is influenced by a nuclear force much stronger than the electrostatic attraction.

As it passes by the nucleus, it is slowed down and deviated in its course, leaving with reduced kinetic energy in a different direction. This loss in kinetic energy reappears as an x-ray photon. These types of x-rays are called bremsstrahlung radiation, or bremsstrahlung x-rays. Bremsstrahlung is the German word for slowing down or braking; bremsstrahlung radiation can be considered radiation resulting from the braking or projectile electrons by the nucleus.

A projectile electron can lose any amount of its kinetic energy in an interaction with the nucleus of a target atom, and the bremsstrahlung radiation associated with the loss can take on a corresponding range of values. For example, an electron with kinetic energy of 70 keV can lose all, none, or any intermediate level of that kinetic energy in a bremsstrahlung interaction; the bremsstrahlung x-ray produced can have an energy in the range of 0 to 70 keV. This is different from the production of characteristic x-rays that have specific energies.

CONTINUOUS X-RAY SPECTRUM

If it were possible to identify and quantitative the energy contained in each bremsstrahlung photon emitted from an x-ray tube, one would find that these energies extend from that associated with the peak electron energy all the way down to zero. In other words, when an x-ray tube is operated at 70 kVp, bremsstrahlung photons with energies ranging from 0 to 70 keV are emitted. Thus, creating a typical continuous, or bremsstrahlung, x-ray emission spectrum.

This emission spectrum is sometimes called the continuous emission spectrum because, unlike in the discrete spectrum, the energies of the photons emitted may range anywhere from zero to some maximum value. The general shape of the continuous x-ray spectrum is the same for all x-ray machines. The maximum energy that an x-ray can have is numerically equal to the kVp of operation. The greatest number of x-ray photons is emitted with energy approximately one-third of the maximum photon energy. The number of x-rays emitted decreases rapidly at very low photon energies and below 5 keV nearly reaches zero.

CHARACTERISTICS AND PRODUCTION OF X-RAYS

Characteristics of X-Rays

- X-rays are invisible.
- X-rays are electrically neutral. They have neither a positive nor a negative charge. They cannot be

accelerated or made to change direction by a magnet or electrical field.

- X-rays have no mass.
- X-rays travel at the speed of light in a vacuum.
- X-rays cannot be optically focused.
- X-rays form a polyenergetic or heterogenous beam.
- The x-ray beam used in diagnostic radiography comprises many photons that have many different energies.
- X-rays travel in straight lines.
- X-rays can cause some substances to fluoresce.
- X-rays cause chemical changes to occur in radiographic and photographic film.
- X-rays can be absorbed or scattered by tissues in the human body.
- X-rays can produce secondary radiation.
- X-rays can cause chemical and biologic damage to living tissue.

X-ray Production

The production of x-rays requires a rapidly moving stream of electrons that are suddenly decelerated or stopped. The source of electrons is the cathode, or negative electrode. Electrons are stopped or decelerated by the anode, or positive electrode. Electrons move between the cathode and the anode because there is a potential difference in charge between the electrodes.

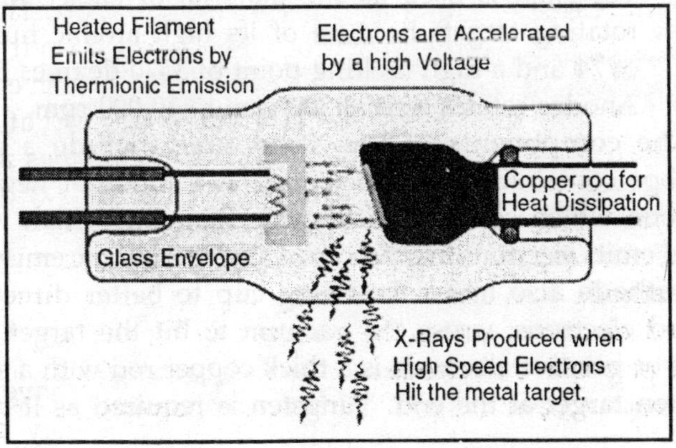

Cathode

- Negatively charged electrode
- Consists of a filament and a focusing cup. Filament is a coiled tungsten wire that serves as the source of electrons during x-ray production.
- Most x-ray tubes are referred to as dual-focus tubes because they use two filaments; a large and a small. Only one filament is energized at any one time during x-ray production.
- Focusing cup is made of nickel and mostly surrounds the filament to focus the stream of electrons before they strike the anode.

Anode

- Positively charged electrode
- Consists of a target and in rotating anode tubes, a stator and rotor.
- The target stops the electrons and creates the opportunity for the production of x-rays.
- The target of rotating anode tubes is made of tungsten and rhenium alloy.
- Tungsten generally makes up 90% of the composition of the rotating target, with rhenium making up the other 10%.
- Rotating anodes generally have a target angle ranging from 6 to 20 degrees.\
- Tungsten is used as the material of choice for the rotating targets because of its high atomic number of 74 and a high melting point of 3370 degrees F.
- Anodes rotates from 3,300 rpm to 10,000 rpm.

The components of the X-ray tube include a glass envelope containing a high vacuum. A cathode or negative electrode which contains a tungsten filament, which when heated emits electrons in a process called'thermionic emission'. The cathode also has a focussing cup to better direct the emitted electrons across the vacuum to hit the target. The anode or positive electrode is a thick copper rod with a small tungsten target at the end. Tungsten is required as it has a high atomic number to improve the efficiency of bremsstrahlung x-ray production and a high melting point.

There is a good deal of heat generated and hence the need for high melting points and the copper anode is able to conduct heat away effectively. Some machines have an oil based cooling system ported through the anode, while others have a spinning electrode to effectively increase the surface area. Either way, considerable heat needs to be dissipated. A potential difference, or voltage is applied between the cathode and anode.

The tungsten filament (cathode) is heated by an independent current and the thermionically emitted electrons are accelerated across the potential difference to a high velocity before striking the tungsten target. The high vacuum is needed to reduce the electron/atom collisions which waste accelerating energy. The electrons that hit the tungsten target undergo sudden deflection because of the interactions with the tungsten nucleus. The tungsten target is usually angled to direct the resultant xrays towards a consistent portion or window in the tube wall.

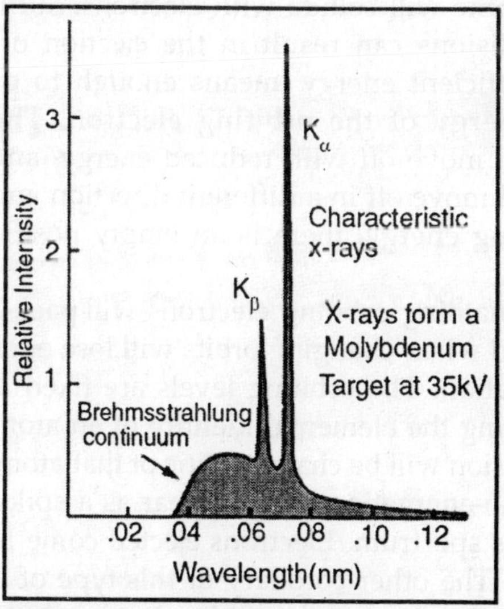

Some additions to this basic set up include the anode hood made of copper and tungsten that act like blinkers to prevent stray electrons from striking the walls of the tube. The copper catches the electrons and the tungsten attenuates the photons produced in the copper. The window is thin and made of beryllium. Beryllium is chosen because it is a metal which has little effect on the photon beam and can effectively maintain the vacuum.

X-Ray Production

X-rays are produced by two main mechanisms and come in two varieties - characteristic and bremsstrahlung xrays

CHARACTERISTIC X-RAYS

Electrons are the same whether orbiting in shells around the nucleus, or produced inside an xray generator. When ever their velocity or position is changed, there is a loss of energy that takes a radiative form (xrays). When electrons travelling at the target have their direction changed, a spectrum of xrays results. However the electrons circulating in the atoms can also change.

While most of the electrons have their path changed and little else, some will collide with electrons. Sufficient energy in such collisions can result in the ejection of an orbiting electron.'Sufficient energy' means enough to overcome the bonding energy of the orbiting electron. The impacting electron will move off with reduced energy, and the ejected electron will move off in a different direction and speed with the remaining energy, there is an empty position in one of the shells.

The remaining orbiting electrons will'pack down' to fill the hole, and when changing orbits will lose energy and emit this as radiation. The orbiting levels are fixed as a physical property fixing the elemental identity of an atom, and so the energy emission will be characteristic of that atom. The energy will be mono-energetic and so appear as a spike rather than a continuous spectrum. Electrons ejected come from the K, L or M orbits. The other corollary of this type of interaction is that the atom becomes an ion (it has lost an ejected electron!).

All atoms will produce characteristic radiation but not

all are visible in the xray portion of the electromagnetic spectrum. Elements with higher atomic numbers have their K, L, M or N shells of sufficient energy to be called'xrays'. The discrete characteristic radiation energies are equal to the difference in the energy level of the outer and inner orbital electrons.

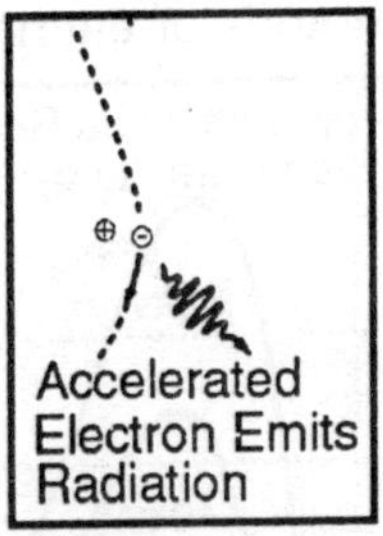

The xray energy is proportional to the atom's Z. Where the incident electrons have energies less than the electron binding energy, there will be no characteristic radiation emitted. As the electron energy increases past the threshold level, the maximum level of characteristic radiation reaches 20% of total production, and then starts to fall to 10% in the 50-100 keV range and 3% in the 200 keV range. In the megavoltage range, characteristic radiation is negligible.

BREMSSTRAHLUNG XRAYS

Bremsstrahlung is a German word meaning"braking radiation" which describes the process of xray generation. The high speed electron impacts on the target and at the atomic level approaches the nucleus. There is no actual collision between electron and nucleus because the electron interacts with the Coulombic nuclear forces and its vector quantities of direction and velocity are changed. Since kinetic energy derives from velocity (KE=1/2mv^2). The change in energy is radiated as electromagnetic radiation. The amount of energy means a short wavelength within the xray band.

As the electron is not destroyed, it can undergo multiple interactions, and even initial interactions will vary from minor to major energy changes depending on the actual angle and proximity of attack, and the point of'impact' on the nucleus.

As a result, bremsstrahlung radiation will have continuous spectrum where the maximum energy relates to the entire KE of the electron but will be infrequent. The energy spectrum without filtration is a straight line that matches the formula,

$$I_E = kZ(E_m - E)$$

where:

IE = intensity of photons of energy E k is a constant,

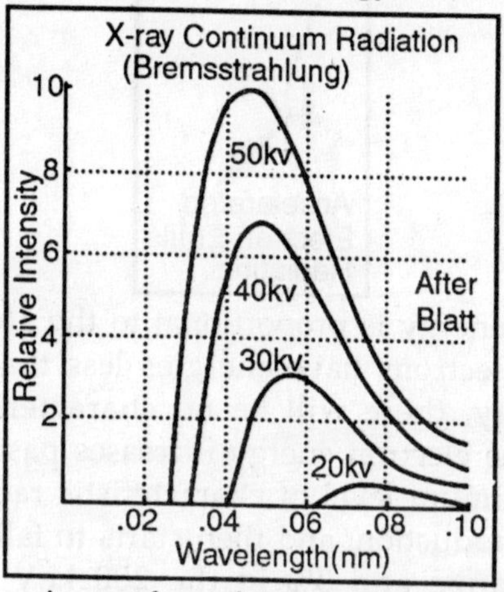

Z is the atomic number of the target, Em is the maximum photon energy which is numerically equal to the applied kilovolts peak (kVp). The average energy of a bremsstrahlung-derived beam is approximately 1/3 of the maximum energy (or kVp). The direction of bremsstrahlung xrays is decidedly horticultural. Where the energy (kVp) of the incident electron beam is around 100 keV, bremsstrahlung production has a spatial orientation described as'anisotropic', that is equally in all directions. The probability of bremsstrahlung production is proportional to the Z^2 of the target material, and the efficiency of production is proportional to the target material Z and the tube voltage.

CHARACTERISTIC AND CONTINUOUS SPECTRA

The spectrum of incandescent solids is said to be continuous because all wavelengths are present. The spectrum of incandescent gases, on the other hand, is called a line spectrum because only a few wavelengths are emitted. These wavelengths appear to be a series of parallel lines because a slit is used as the light-imaging device

CONTINUOUS SPECTRA OF ELECTROMAGNETIC RADIATION

Such spectra are emitted by any warm substance. Heat is the irregular motion of electrons, atoms, and molecules; the higher the temperature, the more rapid is the motion. Since electrons are much lighter than atoms, irregular thermal motion produces irregular oscillatory charge motion, which reflects a continuous spectrum of frequencies. Each oscillation at a particular frequency can be considered a tiny"antenna" that emits and receives electromagnetic radiation.

As a piece of iron is heated to increasingly high temperatures, it first glows red, then yellow, and finally white. In short, all the colours of the visible spectrum are represented. Even before the iron begins to glow red, one can feel the emission of infrared waves by the heat sensation on the skin. A white-hot piece of iron also emits ultraviolet radiation, which can be detected by aphotographic film.

Not all materials heated to the same temperature emit the same amount and spectral distribution of electromagnetic waves. For example, a piece of glass heated next to iron looks nearly colourless, but it feels hotter to the skin (it emits more infrared rays) than does the iron. This observation illustrates the rule of reciprocity: a body radiates strongly at those frequencies that it is able to absorb, because for both processes it needs the tiny antennas of that range of frequencies.

Glass is transparent in the visible range of light because it lacks possible electronic absorption at these particular frequencies. As a consequence, glass cannot glow red because it cannot absorb red. On the other hand, glass is a better emitter/absorber in the infrared than iron or any other metal that strongly reflects such lower frequency electromagnetic waves. This selective emissivity and absorptivity is important for understanding the greenhouse effect

ELECTROMAGNETI SPECTRUM

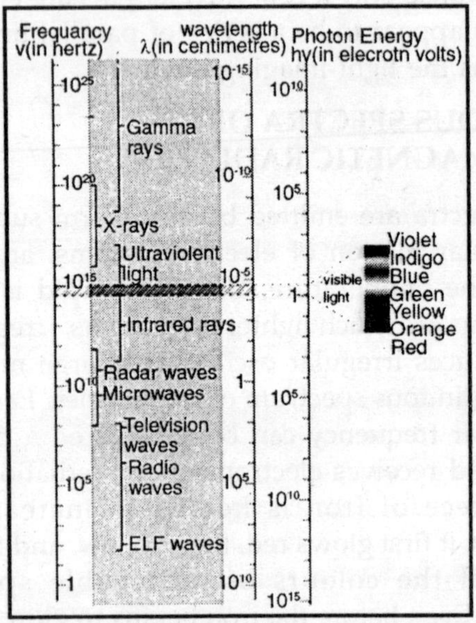

The entire distribution of electromagnetic radiationaccording to frequency or wavelength. Allelectromagnetic waves travel with the same velocity in a vacuum-at the speed of light, which is 299,792,458 metres, or about 186,282 miles, per second. However, the entire distribution covers a wide range of frequencies and wavelengths, and it consists of many subranges, commonly referred to as portions of the electromagnetic spectrum.

The various portions bear different names based on differences in behaviour in the emission, transmission, and absorption of the corresponding waves and also based on their differentpractical applications. There are no precise accepted boundaries between any of these contiguous portions, so the ranges tend to overlap.

The entire electromagnetic spectrum, from the lowest to the highest frequency (longest to shortest wavelength), includes all radio waves (e.g., commercial radio and television, microwaves, radar), infrared radiation, visible light, ultraviolet radiation, X-rays, and gamma rays.

Chapter 4

Nuclear Instability

NUCLEAR INSTABILITY

NUCLEUS AND NUCLEAR INSTABILITY

Nuclei are described using the following nomenclature:

$$^{A}_{Z}\text{Element}_{N}$$

Z is the atomic number, the number of protons: this defines the element.

- A is called the"mass number" A = N + Z.
- N is the number of neutrons (N = A – Z)
- *Nuclide*: A species of nucleus of a given Z and A.
- *Isotope*: Nuclides of an element (i.e. same Z) with diffcrent N.
- *Isotone*: Nuclides having the same N.
- *Isobar*: Nuclides having the same A.

A handy way to keep these straight is to note that isotope includes the letter"p" (same proton number), isotone the letter"n" (same neutron number), and isobar the letter"a" (same A).

Example:

$$^{206}_{82}\text{pb}_{124}$$

of isotope an is,

$$^{208}_{82}\text{pb}_{126}$$

of isotone an is,

$$^{207}_{83}\text{Bi}_{124}$$

of isobar an and,

$$^{206}_{84}\text{Po}_{122}$$

90 natural elements

109 total elements

All elements with Z > 42 are man-made Except for technetium Z=43 Promethium Z = 61. More than 800 nuclides are known (274 are stable)"stable" unable to transform into another configuration without the addition of outside energy."unstable" = radioactive

NUCLEAR STRUCTURE: FORCES

Coulomb Force

Force between two point charges, q, separated by distance, r (Coulomb's Law).

$$F(N) = \frac{k_o q_1 q_2}{R^2} k0 = 8.98755 \times 10^9 \, N\, m^2 C^{-2} \, (\text{Boltzman cons tan t})$$

Potential energy (MeV) of one particle relative to the other:

$$PE(MeV) = \frac{k_0 q_1 q_2}{R}$$

Nuclear Force

Summary of Nuclear Forces

Nuclei give off energy (i.e., radiation) in an attempt to become more stable. Nuclear instability can be traced to the interaction of i) Coulomb and ii) strong nuclear force.

Coulomb	Strong Nuclear
Repulsive	Attractive
p+ - p+	p+- p+, n - n, p+- n
doesn't saturate	short range; falls off quickly
weak (eg. e- to nucleus, MeV)	ery strong (several decades of
Atom is mostly empty space	nucleus is densely packed

Due to the Coulomb-nuclear force balance, nuclei exhibit a roughly constant density and radius.

ENERGY-MASS EQUIVALENCE

Atomic Mass Units (amu, or AMU)

By definition: Atomic masses are measured on a scale in

which a $^{12}C_6$ atom is exactly 12 atomic mass units. Gram atomic weight of any element contains N_0 atoms (N_0 = Avogadro's number) 12 grams of carbon = 6.02×10^{23} carbon atoms,

$$\frac{12\text{gcarbon}}{6.02\times10^{23}\text{ atoms}}=\frac{1.99\times10^{-23}\text{g}}{\text{carbon atom}}=12\text{AMU}$$

$$1\text{aum}=\frac{1.99\times10^{-23}\text{g}}{}=1.66\times10^{-24}\text{g}=1.66\times10^{-27}\text{g}$$

Using Einstein's mass-energy equivalence formula: $E=m_0C^2$,

$1.660531 \times 10\text{-}^{27}$ kg × (3.0×10^8 m/s) = 1.49448 × 10-10 kg m2/s2 = 1.49448 × 10-^{10} Joule

Given: 1.6022 × 10-^{19} Joules = 1 eV

1 AMU is equivalent to 931.48 MeV

Rest mass energies and mass equivalences:

- *Electron mass*: 0.000549 amu = 0.511 MeV
- *Proton mass*: 1.007277 amu = 938.28 MeV
- *Neutron mass*: 1.008665 amu = 939.57 MeV
- *Hydrogen atom*: 1.007825 amu

Mass Differences

The mass of a nuclide is LESS than the sum of its parts:

Energy released when all constituents come together.

Nuclear force so strong that the mass of the bound system is smaller than the *sum of the components*:

$$\Delta = M - A, \text{ or } M = \Delta + A$$

M is the true atomic mass

A is the atomic number

NUCLEAR REACTIONS RELEASE ENERGY

$${}_0^1n + {}_1^1H \rightarrow {}_1^2H + {}_0^0\gamma$$

How much energy is released?

Compare the total masses on both sides of the arrow.

NUCLEAR BINDING ENERGIES

The difference in mass between a given nucleus and the

sum of the same number of individual protons and neutrons is the binding energy.

Nuclear Stability/Instability

Strong nuclear force, operates over short range is"saturates" quickly neutrons interact only with neighbours protons interact (repulse via Coulomb interaction) throughout the entire nucleus.

In heavier nuclei, the #neutrons must increase faster than the number of protons to maintain stability.

N/Z ratio = 1 at low A

Mg Z=12, but N=12,13 or 14 (isotopes)

N/Z ratio approaches 1.5 when Z~80

"Line of stability" Z = N

Any nucleus far from the"line of stability" will be unstable. The position of a nucleus relative to the line of stability will define the mode of nuclear instability (radioactive decay mode).

Alpha decay

Natural alpha emitters: Z > 83

$$^{A}_{Z}P \rightarrow ^{A-4}_{Z-2}D + ^{4}_{2}He + Q$$

Conservation of electric charges.

Conservation of nucleons,

$$\text{E.g., } ^{226}_{88}Ra \rightarrow ^{222}_{86}Rn + ^{4}_{2}He$$

How much energy, Q, is released?

Compare the masses on both sides of the arrow.

Q = MRa – MRn – Mhe

Use. values in Turner Appendix D.

$Q = \Delta_{Ra} - \Delta_{Rn} - \Delta_{He}$

Q = 23.69 – 16.39 – 2.42 = 4.88 MeV

How is this energy, Q, distributed?

Shared by the daughters, the Rn nucleus and the alpha particle.

Momentum is conserved: mv = MV

Kinetic energy of the 2 products = Q

$$\frac{1}{2}mv^2 + \frac{1}{2}MV^2 = Q$$

The energy of the alpha particle:

$$E_\alpha = \frac{1}{2}mv^2 = \frac{MQ}{m+M}$$

The energy of the Rn nucleus:

$$E_N = \frac{1}{2}MV^2 = \frac{mQ}{m+M}$$

Alpha decay results in a 2-particle emission.
Q is fixed by the mass balance
Eα is fixed by the conservation laws (energy, momentum)
Therefore, alpha particles must have discrete energies.

Nuclear Decay Scheme Diagrams

Graphical display of nuclear transformations:

- Decay mode
- Energy transitions
- Abundances (branching ratios)

Conventions:

- Arrows slanting to the left indicate decrease in Z
- Arrows slanting to the right indicate an increase in Z
- Wavy lines going straight down indicate a gamma emission from the nucleus.

Image removed.
Appendix D in [Turner]

Decay Scheme Exercise:

$$^{219}_{86}Rn \rightarrow ^{215}_{82}Po + ^{4}_{2}He$$

α 6.82 (80%)
6.55 (12%)
6.42 (7%)
γ 0.271 (10%)
0.402 (7%)

UNSTABLE NUCLEAR POWER

Stable power processes are never guaranteed. An

assortment of unstable behaviours wrecks power apparatus, including mechanical vibration, malfunctioning control apparatus, unstable fluid flow, unstable boiling of liquids, or combinations thereof.

INHERENTLY SAFE" REACTORS

The experts tell us that they will build"safe" reactors, using inherent reactor characteristics to limit any unwanted rise in reactor power. A rise in power increases the temperature of reactor parts, causing them to swell, or bend, or otherwise distort. A necessary condition for stable reactor power stable is that the reactor distortion accompanying a temperature rise decrease the reactivity, thus slowing any further rise in power.

For instance, the reactor distortion caused by a temperature rise could allow more neutrons to be wasted by leakage, thus decreasing the reactivity. Overly simple equations for reactor power dynamics indicate that this idea should work, and sometimes it does.

The supposition that satisfying the above condition guarantees power stability is based on a false assumption that the reactor distortion occurs simultaneously with the temperature rise which caused it. The distortion lags behind the temperature change by a time determined by the mechanical dynamics of the reactor system. Therefore the change in reactivity also lags behind the temperature change, raising the possibility of unstable power oscillations. This conclusion is supported by actual experience with seriously unstable reactors such as the inoperable Ft. St. Vrain power reactor discussed later.

It is shown in the last section of this pamphlet that stability against power oscillations depends on having sufficient mechanical friction in the distorting reactor parts. Guaranteeing this requirement may be outside the power of a designer.

EVIDENCE OF REACTOR INSTABILITY

My interest in the subject of reactor instability was

aroused by two sets of experiments. I heard of the first set of experiments, called"teasing the dragon," from a physicist who had worked on the atomic bomb project. In a low-level simulation of a bomb test, a sudden pulse of nuclear energy was initiated, sufficient to cause parts of the test assembly to jump apart momentarily (and occasionally to destroy the experimenter, but not quite enough to destroy the apparatus and its surroundings). I wanted to understand how the dynamic mechanism which controlled the energy pulse might be related to the control of reactor power.

The second set of experiments, under the acronym, SPERT, was carried out at the National Reactor Test Station to study transients of reactor power. An assembly of thin sheets of aluminum bearing the nuclear fuel was immersed in a pool of water, the surface of the water being open to the atmosphere. Reactor power was slowly increased by gradually pulling out control rods.

At a certain low power a spontaneous unstable oscillation suddenly appeared and built rapidly to violent proportions, resulting in a flash of blue light from Cerenkov radiation and the explosive expulsion of water from the reactor pool. The peak of power was greater by a factor of thousands than the level from which the oscillations had started. It was concluded that the unexpected instability was related to coolant boiling.

In the early 1960's I developed an analysis of a possible unstable mode of power oscillation which I thought could lead to reactor explosions. I developed mathematical criteria to try to predict the threshold of reactor power at which such unstable oscillations would occur. Based on my analyses, I concluded that coolant boiling in the SPERT reactor set up a dynamic configuration in which a heavy mass of water bounced at a low natural frequency on the soft spring provided by vapour bubbles.

The mechanical oscillation was coupled with the nuclear power oscillation in an unstable combination. Water-moderated reactors, the type most commonly used for the generation of power, might be susceptible to catastrophic nuclear power oscillations if the pressure of the water in the reactor suddenly decreased, creating vapour bubbles in the core. I wrote several reports predicting that various power reactors could be susceptible to power oscillations of the type observed in the SPERT experiments. One preliminary report entitled"Study of Reactor Kinetics" was published by the American Society of Mechanical Engineers as Paper Number 62-WA-218 late in 1961. I presented the report at the annual meeting of the Society in New York in November 1962. I received encouragement from engineering friends, but was unable to get any attention from the nuclear establishment. I was denied publication in the nuclear journals.

The internal study and public scrutiny of potential instabilities of reactor power was (and is) not popular with reactor proponents. Even when instabilities cause accidents, they are not acknowledged as such.

UNSTABLE REACTORS

The SPERT experiments seemed to us pertinent to our claim that mechanical effects contribute to reactor instability. Subsequent reactors have demonstrated instabilities similar to our predictions.

A non-nuclear model of a nuclear reactor, for a space propulsion project named ROVER, was composed of large graphite blocks through which a gas was circulated in a flow test. The blocks were restrained against the radial pressure of the gas by springs at their outer periphery. A motion picture taken during a flow test showed the blocks banging to pieces as the pressure of the circulating gas bounced them against the restraining springs in a self-excited oscillation, equivalent to a negative mechanical damping in the system. No nuclear catastrophe occurred only because there was no nuclear power to couple with the mechanical oscillation of the graphite blocks.

A high-temperature, gas-cooled reactor for electrical power was built for Colourado Public Service at Ft. St. Vrain. On completion it could not be brought to a useful level of operating power because persistent power oscillations worsened as power was increased. Because of the oscillation and other problems the reactor was abandoned. A successful ratepayers' suit removed the useless power plant from the rate base.

REACTOR KINETICS

Twenty-five years after my failure to achieve publication in nuclear journals, my son, Bruce Thompson, volunteered to help me write a straight technical article about stability, based on a computer programme which we had developed together. Our programme solves by finite difference mathematics a non-linear, fourth-order differential equation involving nuclear, mechanical and thermal characteristics of reactors. Our paper,"A Model of Reactor Kinetics," was finally published in Nuclear Science and Engineering, the technical journal of the American Nuclear Society, in September 1988. It demonstrates a mechanism for catastrophic instability.

Our paper demonstrates that mechanical friction in a reactor core structure, like the shock absorber in an automobile, is necessary to limit oscillations of reactor power. Without adequate internal friction, a nuclear power driven mechanical oscillation increases toward destruction of the reactor core. Design engineers in many fields have found to their sorrow that any given level of mechanical friction is difficult to guarantee. Some of the computer-generated examples later in this section show the changing core temperature and reactor power for a reactor without adequate friction to provide stability. A small perturbation in power causes an initially small oscillation which builds rapidly to destruction, either blowup or melt down.

In the design, construction and operation of nuclear reactors an attempt is made to maintain steady operation at any desired power level from fission of the nuclear fuels, uranium or plutonium. This attempt can fail, sometimes catastrophically. A nuclear power plant is a nuclear system and a mechanical system. It is also a heat-transfer system, tied in with controls, boilers, turbines, human operators and a multiplicity of other complicating factors.

The possibilities for instability are myriad. This fiercely

complicated set of systems is such as to preclude any possibility of the formation of adequate analytical equations and their solution to guarantee the stability of power. An experimental programme sufficient to eliminate the possibilities of power instability in reactors would be expected to be ruinous of both the economy and the environment. Our programme covers only a small part of the complicated possibilities, and demonstrates only one type of power instability.

PERTURBATIONS TO REACTOR POWER

We outline here the calculation of the power dynamics of a reactor, based on. We present some necessary conditions for power stability provided by nuclear physics, as analysed by Weinberg and Wigner in.

Their analysis generated a second-order differential equation whose solution is always stable if the coefficients in the equation are all positive. Then we consider mechanical characteristics of reactors which render the nuclear analysis insufficient to assure stability.

This results in a fourth-order differential equation whose stability requires not only that all coefficients be positive, but also that a relationship among the coefficients be satisfied (Routh's criterion). It turns out that, for power stability, motion of any reactor parts which affect reactivity must be sufficiently damped by mechanical friction in the moving parts.

Reactivity and stability are unavoidably affected by depletion of reactor fuels and the accumulation of"poisons" during operation. Other destabilizing effects not considered here, or anywhere else, are so numerous and so complicated in their interrelationships as to thwart human efforts to guarantee reactor stability.

Example

We attempt with twelve examples to demonstrate some physical properties of reactors on which stability depends. The examples show the influence on power stability of:

reactivity, delayed neutrons, coefficients of reactivity, mechanical inertia, friction, heat transfer.

The defining parameters for these examples are:

A: Time interval
B: Final time
C: Frequency ratio, density coefficient
D: Frequency ratio, temperature coefficient
E: Mechanical damping
F: Nuclear damping
G: Initial reactivity
H: Initial power ratio
I: Mechanical effects on (off)
J: Heat transfer on (off)
K Coefficient of Heat Transfer

The values used for A through K, are shown with each of the computer-generated solutions.

Nuclear and Mechanical Characteristics

The first examples show the effects of the"in-built" nuclear characteristics, delayed neutrons and temperature coefficients of reactivity, which reactor proponents say will"guarantee" the stability of their new generation of"safe" reactors. The later examples show some of the destabilizing mechanical effects which may defeat both the proponents of reactors and the rest of us.

Chapter 5

Introduction to Nuclear Physics

INTRODUCTION

REACTIVITY

Reactivity is the name of the quantity which determines the rate of change of reactor power. For reactor power to be steady the reactivity must be zero.

Maintaining zero reactivity requires moving"control rods" to maintain a balance among the rates of leakage and absorption of neutrons in nuclear fuel and in other reactor materials. Because the absorption and leakage of neutrons have been extensively studied, these balancing dependencies are relatively well understood. Positive reactivity causes power to rise exponentially at a rate proportional to the reactivity. Negative reactivity causes power to decrease.

To change power in a planned manner, reactivity is adjusted by moving the control rods, either manually or by means of automatic controls. Partially removing a control rod is expected to increase the reactivity, causing the power to rise to a new level. Inserting a properly designed control rod farther into the reactor should, but doesn't always, lower the power level.

Example shows the effect of a relatively large initial reactivity, ignoring several important limiting effects, leading to a rapid rise to a"blowup." The unacceptable behaviour which would result from a continuation of this exponential response is modified in real reactors by two important effects.

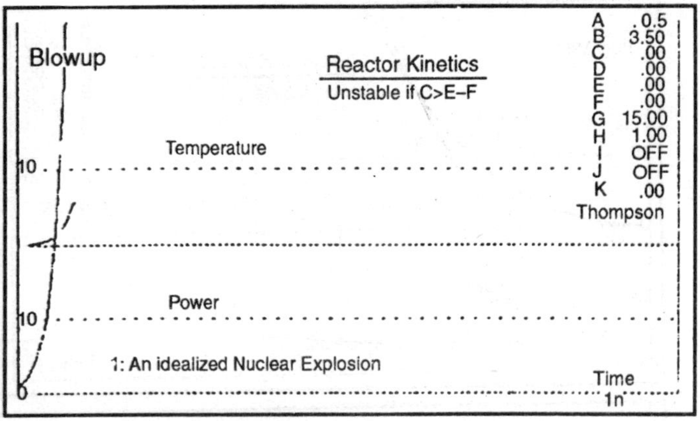

DELAYED NEUTRONS

The first nuclear effect is that of"delayed neutrons." Of all neutrons formed in fission a small fraction (.007) are delayed by significant time periods after absorption of the fission neutrons.

The delayed neutrons, not being available immediately to support a transient rise in power, act as a damping factor on transients, providing the strong beneficial effect of limiting sudden changes in power.

Example below show the classical"damping effect" of delayed neutrons on the rise of power, ignoring the temperature coefficient of reactivity and mechanical effects. Example 3 shows the reaction to"prompt critical" reactivity, in which by definition the initial reactivity is numerically equal to the fraction of delayed neutrons (G=F=15).

The initial rate of power rise continues indefinitely on a straight line. In Example 2the addition of reactivity greater than prompt critical (G/F = 15/14) causes an exponential rise of power faster than in Example 3. Example 4 shows a"sub-critical" reaction, in which a value of reactivity less than the fraction of delayed neutrons,

(G/F = 15/16)

leads to the leveling-off of power at a new value,

P/P0=F/(G-F) = 16.

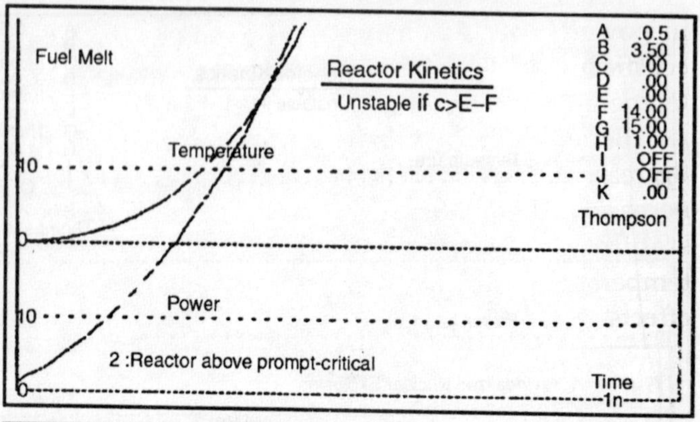

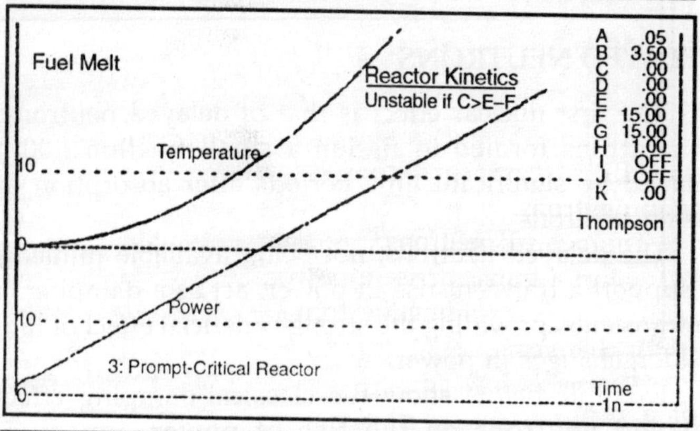

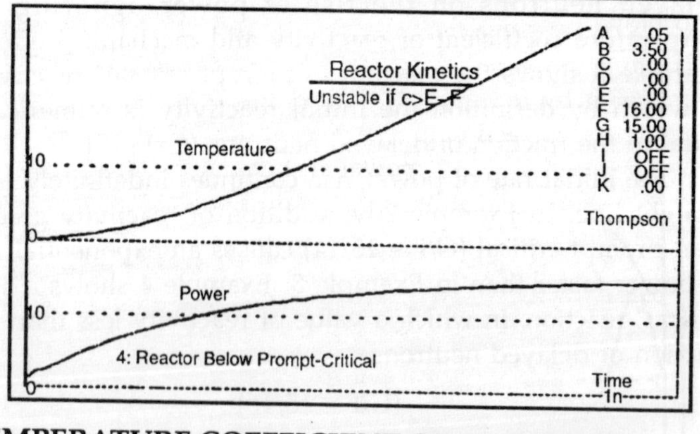

TEMPERATURE COEFFICIENT OF REACTIVITY

The rise in temperature which accompanies a rapid rise

in power generally changes the reactivity. The"temperature coefficient of reactivity" measures the effect of changing temperature on reactivity.

Because power stability requires a rise in temperature to decrease the reactivity, a"negative temperature coefficient" is necessary for stability. Examples show the effects on power stability of including delayed neutron damping and temperature coefficients of reactivity, but neglecting all other effects.

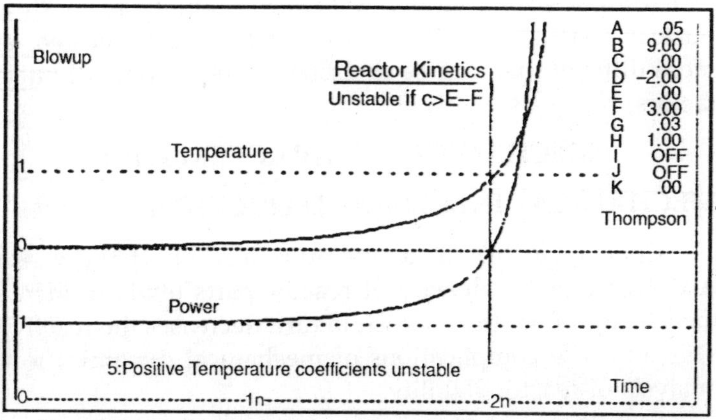

Example shows the destabilizing effect on reactor power of a positive temperature coefficient (D= −2.0), leading to"blowup."

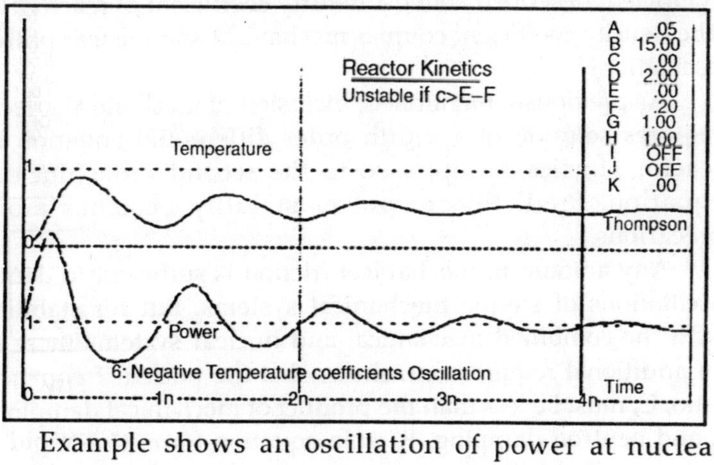

Example shows an oscillation of power at nuclear

frequency, with a negative temperature coefficient (D=2.0) and with delayed neutron damping (F=0.20). Because of neutron damping the oscillations decrease in magnitude with time.

The effects of the nuclear factors, reactivity, delayed neutrons and temperature coefficients, are described by a non-linear second-order differential equation.

This relatively simple equation predicts stable reactor power if the temperature coefficient of reactivity is negative. Neutron damping in reactors is generally relatively large, providing a strong tendency toward stability for a simple nuclear system. The real situation is more complicated, and more difficult than can be described in terms only of nuclear factors.

MECHANICAL CHARACTERISTICS

MECHANICAL INERTIA AND FRICTION

Changes in reactor temperature also affect the density. A decrease in the density of reactor parts tends to increase the leakage of neutrons which in turn decreases the reactivity. This adds the complications of mechanical dynamics to the analysis of reactor stability.

Fast density changes in a reactor structure may cause it to vibrate, and reactor power may oscillate as a result. Accounting for this effect requires adding to the temperature coefficient described above a"density coefficient of reactivity." The density coefficient couples mechanical and nuclear power oscillations.

As previously mentioned, inclusion of mechanical effects requires solution of a fourth order differential equation of reactor kinetics, as opposed to the second order nuclear equation. Prediction of power stability becomes more precarious.

Any amount of mechanical friction is sufficient to damp oscillations of simple mechanical systems. But for stability with the combined mechanical and nuclear systems there is an additional requirement. shows that the nuclear frequency ratio, C, must be less than the product of mechanical damping, E and neutron damping, F, or in equation form, $C<E*F$.

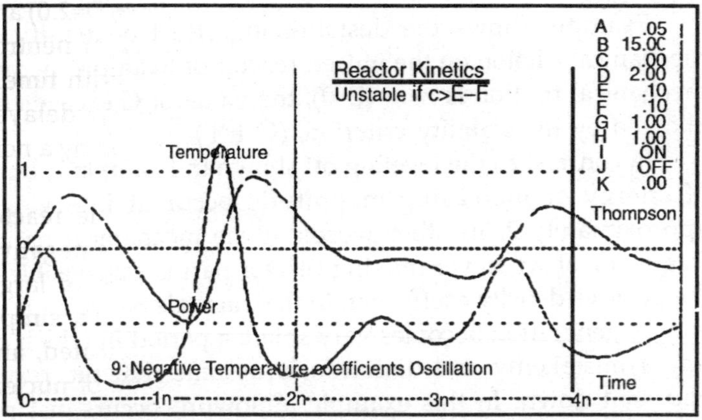

Example Models a stable pulsed reactor, like those used in nuclear engineering departments at universities, and in some bomb tests.

When the control rod is suddenly withdrawn, the large reactivity (G=12), above the prompt critical set by delayed neutrons (F=7), causes the temperature to overshoot the equilibrium value (G/C=6), and the power suddenly rises. Because the product of the mechanical friction (E=.4) and delayed neutron damping (F=7) is greater than the frequency ratio (C=2) as required by the stability criterion, E*F>C, the configuration is stable. After its initial burst power finally levels off near its steady state value, P/P0=1.

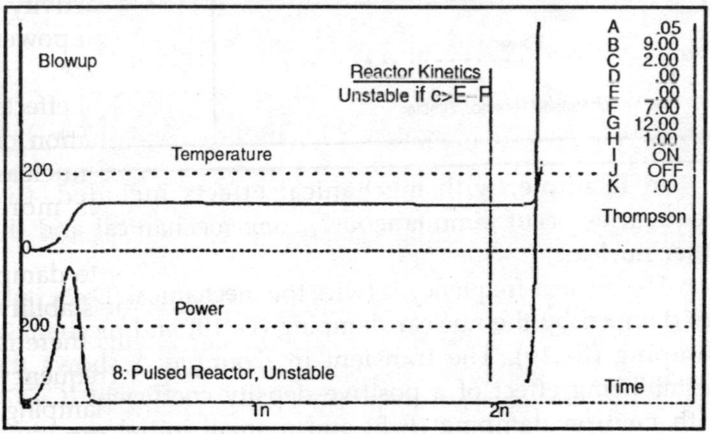

Example shows the destabilizing effect of insufficient mechanical friction on the pulsed reactor of Example. Because mechanical friction is zero (E=0), the value of C exceeds that allowed by the stability criterion, (C>E*F).

In contrast to the leveling off shown in Example 7, bursts of energy of increasing magnitude occur at intervals of approximately 2, in other words at the mechanical natural frequency of whatever moving reactor part is responsible for the positive density coefficient. In the intervals between peaks, energy generation becomes very small, a period in which one might mistakenly and disastrously assume that the reactor had shut down. In this example a"blowup" occurs just past the end of the first mechanical cycle.

In Examples through C the scale has been expanded by a large factor to accommodate smaller transients appropriate to the operation of power reactors.

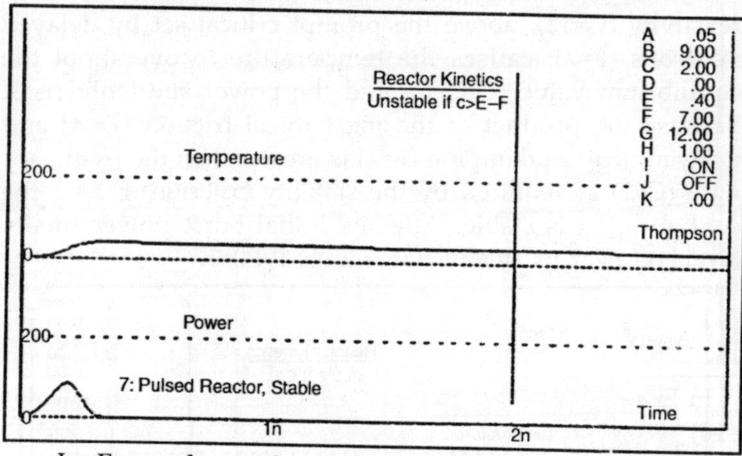

In Example, with mechanical effects included, two frequencies occur simultaneously, one mechanical and the other nuclear.

The nuclear frequency is twice the mechanical (D=2). Both are damped by the nuclear damping (F=.10) and mechanical damping (E=.10). The transient in Example A shows the destabilizing effect of a positive density coefficient (C=20), with neutron damping (F=5) and a small initial reactivity (G=.05), but with inadequate mechanical damping (E=.05).

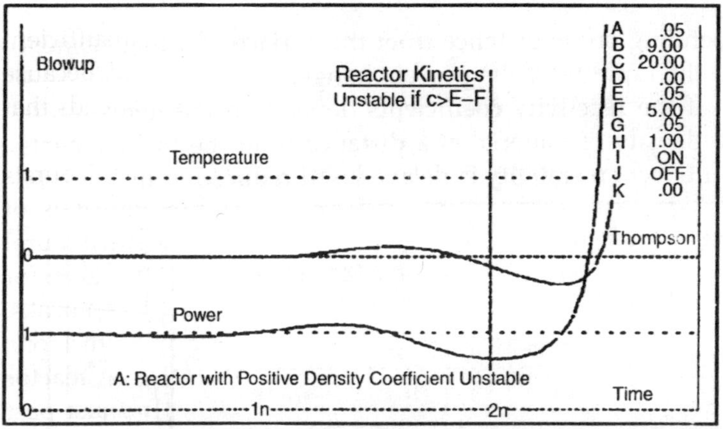

This, as expected from stability analysis (C>E*F), causes power bursts at approximately mechanical frequency of ever increasing magnitude.

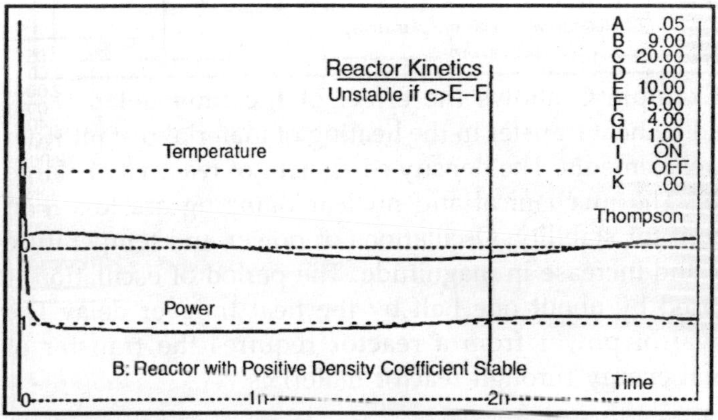

In Example B addition of sufficient mechanical damping (E=10) returns the system to stability (C<E*F), despite the larger initial reactivity (G=4).

HEAT TRANSFER DELAYS THE TEMPERATURE RESPONSE

When power generation in a reactor changes rapidly with time, temperature waves are propagated from fuel elements into the surroundings. Because of the time required for thermal energy to travel through materials, the temperature change at any point is delayed and reduced in magnitude depending on its distance from the surface of a fuel element and the rapidity of the power change.

If the reactivity coefficients depend on the temperature and density of material at a distance from the fuel elements their effect on stability is delayed and reduced in magnitude.

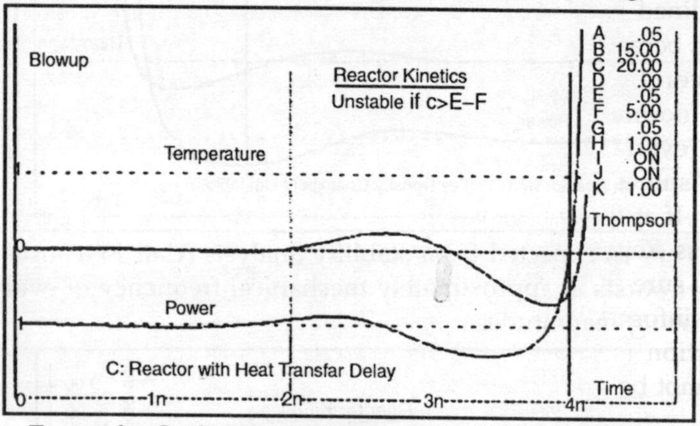

Example C shows the effect of the time delay (K=1) caused by heat transfer in the heating of material distant from the fuel elements. The density coefficient of reactivity is large (C=20). The mechanical and nuclear damping are less than required for stability. Oscillations of power and temperature occur and increase in magnitude. The period of oscillation is increased by about one half by the heat transfer delay.The removal of power from a reactor requires the transfer of thermal energy through reactor materials.

Conclusions

We have discussed here some effects on stability of mechanical dynamics and heat transfer, in addition to nuclear effects. We believe that this study of Reactor Kinetics may have pertinence to understanding, among other events, the Spert and Rover experiments, the oscillating power in the Ft. St. Vrain reactor, and some aspects of the accidents to SL-1.

Within the complicated systems of a nuclear power plant a multiplicity of instabilities other than those we have treated is probable. We believe a complete analysis of stability applicable to a nuclear power plant would require the inclusion of other effects too numerous and complicated for ordinary humans to handle with certainty.

Is our analysis pertinent?

So long as reactors are built, we believe the effects on stability of mechanical friction, inertia, and heat transfer outlined here should be taken into account in their design. But, because of the complication of nuclear reactors, they can never be guaranteed free from accidents. Accidents are"normal" but reactor accidents are intolerable. Because of the everlasting radioactive products it manufactures, even the"successful" reactor is intolerable.

If at the A:> prompt on my computer I type DEL *.* and press Return there appears on the screen the question, Are you sure (Y/N). Because * is a wild-card, I am warned that all the information on the disk is at risk. We believe mechanical friction is a wild-card of nuclear reactors whose efficacy cannot be guaranteed by reactor designers.

RADIOACTIVE DECAY LAWS

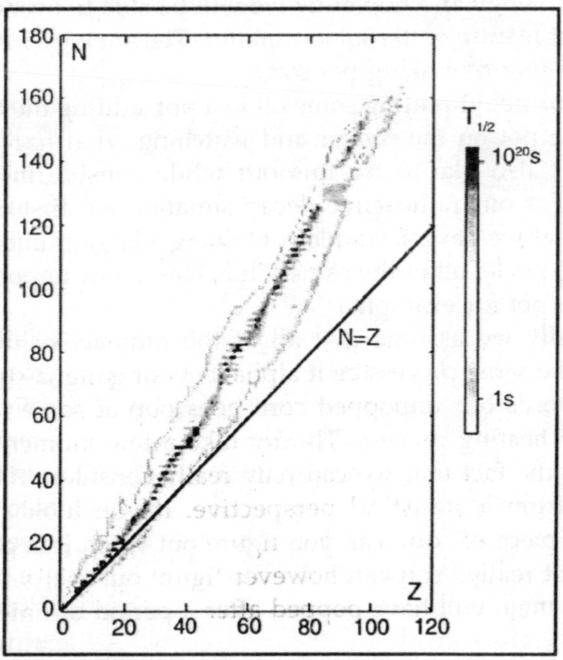

We covered radioactive decay from a phenomenological perspective in the last chapter. In this chapter we consider the topic from a more general analytical perspective.

The reason for doing this is so that we can develop a form of thinking which will help us to understand what is going on in a quantitative, mathematical sense. We will be introduced to concepts such as the Decay Constant and the Half Life as well as units used for the measurement of radioactivity.

ASSUMPTIONS

The usual starting point in most forms of analysis in physics is to make some assumptions which simplify the situation. By simplifying the situation we can dispose of irrelevant effects which tend to complicate matters but in doing so we sometimes make the situation so simple that it becomes a bit too abstract and apparently hard to understand.

For this reason we will try here to relate the subject of radioactive decay to a more common situation which we will use as an analogy and hopefully we will be able to overcome the abstract feature of the subject matter. The analogy we will use here is that of making popcorn.

So think about putting some oil in a pot, adding the corn, heating the pot on the cooker and watching what happens. You might also like to try this out while considering the situation! For our radioactive decay situation we first of all consider that we have a sample containing a large number of radioactive nuclei all of the same kind. This is our unpopped corn in the pot for example.

Secondly we assume that all of the radioactive nuclei decay by the same process be it alpha, beta or gamma-decay. In other words our unpopped corn goes pop at some stage during the heating process. Thirdly take a few moments to ponder on the fact that we can only really consider what is going on from a statistical perspective. If you look at an individual piece of corn, can you figure out when it is going to pop? Not really. You can however figure out that a large number of them will have popped after a period of time.

But its rather more difficult to figure out the situation for an individual piece of corn. So instead of dealing with individual entities we consider what happens on a larger scale and this is where statistics comes in. We can say that the radioactive decay is a statistical one-shot process, that is when a nucleus has decayed it cannot repeat the process again. In other words when a piece of corn has popped it cannot repeat the process. Simple! In addition as long as a radioactive nucleus has not decayed the probability for it doing so in the next moment remains the same. In other words if a piece of corn has not popped at a certain time the chance of it popping in the next second is the same as in the previous second. The bets are even!

Let us not push this popcorn analogy too far though in that we know that we can control the rate of popping by the heat we apply to the pot for example. However as far as our radioactive nuclei are concerned there is nothing we can do to control what is going on.

The rate at which nuclei go pop (or decay, in other words) cannot be influenced by heating up the sample. Nor by cooling it for that matter or by by putting it under greater pressures, by changing the gravitational environment by taking it out into space for instance, or by changing any other aspect of its physical environment. The only thing that determines whether an individual nucleus will decay seems to be the nucleus itself. But on the average we can say that it will decay at some stage.

THE RADIOACTIVE DECAY LAW

Let us now use some symbols to reduce the amount of writing we have to do to describe what is going on and to avail ourselves of some mathematical techniques to simplify the situation even further than we have been able to do so far. Let us say that in the sample of radioactive material there are N nuclei which have not decayed at a certain time, t. So what happens in the next brief period of time? Some nuclei will decay for sure.

But how many?

On the basis of our reasoning above we can say that the number which will decay will depend on overall number of nuclei, N, and also on the length of the brief period of time. In other words the more nuclei there are the more will decay and the longer the time period the more nuclei will decay. Let us denote the number which will have decayed as dN and the small time interval as dt.

So we have reasoned that the number of radioactive nuclei which will decay during the time interval from t to t+dt must be proportional to N and to dt.

In symbols therefore:

$$-dN \propto N.dt$$

the minus sign indicating that N is decreasing.

Turning the proportionality in this equation into an equality we can write:

$$-dN=\lambda N.dt$$

where the constant of proportionality, λ, is called the Decay Constant.

Dividing across by N we can rewrite this equation as:

$$-\frac{dN}{N}=\lambda dt$$

So this equation describes the situation for any brief time interval, dt. To find out what happens for all periods of time we simply add up what happens in each brief time interval. In other words weintegrate the above equation. Expressing this more formally we can say that for the period of time from t = 0 to any later time t, the number of radioactive nuclei will decrease from N_0 to N_t, so that:

$$-\int_{N_0}^{N_t}\frac{dN}{N}=\lambda\int_0^t dt$$

$$\therefore \ln\left(\frac{N_t}{N_0}\right)=-\lambda t$$

$$\therefore \frac{N_t}{N_0}=\exp(-\lambda t)$$

$$\therefore N_t = N_0 \exp(-\lambda t)$$

This final expression is known as the Radioactive Decay Law. It tells us that the number of radioactive nuclei will decrease in an exponential fashion with time with the rate of decrease being controlled by the Decay Constant.

Before looking at this expression in further detail let us review the mathematics which we used above. First of all we used integral calculus to figure out what was happening over a period of time by integrating what we knew would occur in a brief interval of time. Secondly we used a calculus relationship that the,

$$\int \frac{dx}{x} = \ln x$$

where ln x represents the natural logarithm of x. And thirdly we used the definition of logarithms that when

$$\ln x = y$$

then,

$$x = \exp y$$

Now, to return to the Radioactive Decay Law. The Law tells us that the number of radioactive nuclei will decrease with time in an exponential fashion with the rate of decrease being controlled by the Decay Constant. The Law is shown in graphical form in the figure below:

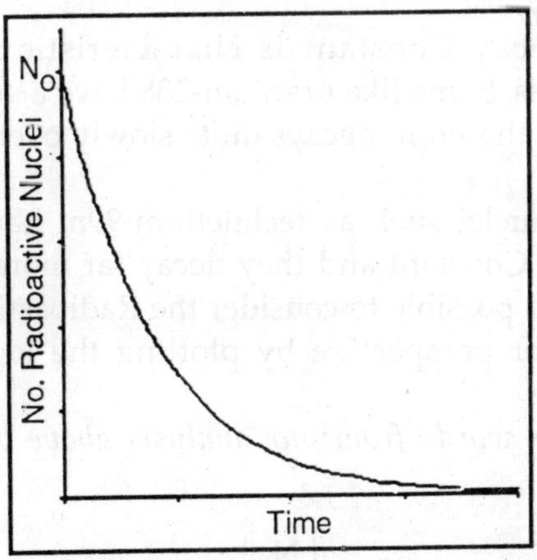

The graph plots the number of radioactive nuclei at any

time, N_t, against time, t. We can see that the number of radioactive nuclei decreases from N_0 that is the number at t = 0 in a rapid fashion initially and then more slowly in the classic exponential manner.

The influence of the Decay Constant can be seen in the following figure:

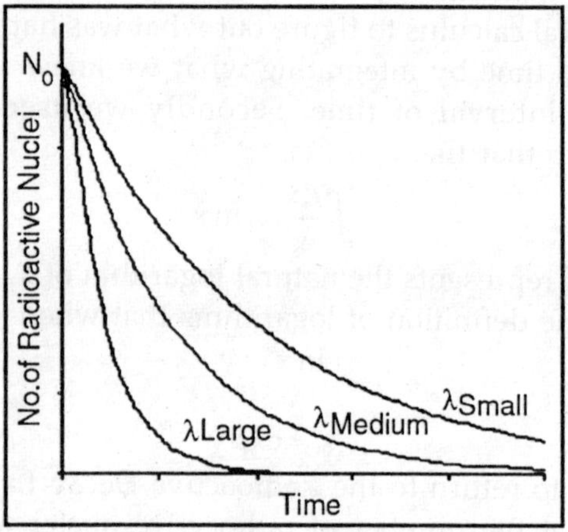

All three curves here are exponential in nature, only the Decay Constant is different. Notice that when the Decay Constant has a low value the curve decreases relatively slowly and when the Decay Constant is large the curve decreases very quickly.

The Decay Constant is characteristic of individual radionuclides. Some like uranium-238 have a small value and the material therefore decays quite slowly over a long period of time.

Other nuclei such as technetium-99m have a relatively large Decay Constant and they decay far more quickly.

It is also possible to consider the Radioactive Decay Law from another perspective by plotting the logarithm of Nt against time.

In other words from our analysis above by plotting the expression:

$$\text{In}\left(\frac{N_t}{N_0}\right) = -\lambda t$$

in the form,

$$\ln N_t = -\lambda + \ln N_0$$

Notice that this expression is simply an equation of the form y = mx + c where m = -l and c = ln N0. As a result it is the equation of a straight line of slope -l as shown in the following figure. Such a plot is sometimes useful when we wish to consider a situation without the complication of the direct exponential behaviour.

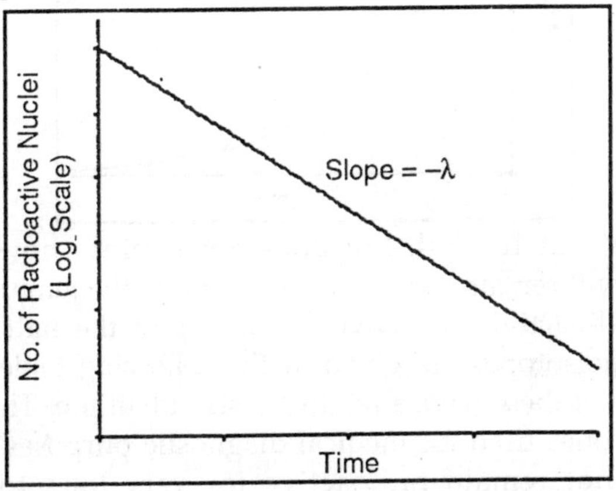

HALF-LIFE

Most of us have not been taught to think instinctively in terms of logarithmic or exponential terms even though many natural phenomena display exponential behaviours.

Most of the forms of thinking which we have been taught in school are based on linear changes and as a result it is rather difficult for us to grasp the Radioactive Decay Law intuitively. For this reason an indicator is usually derived from the law which helps us think more clearly about what is going on.

This indicator is called the Half Life and it expresses the length of time it takes for the radioactivity of a radioisotope to decrease by a factor of two.

From a graphical point of view we can say that when:

$$N_t = \frac{N_0}{2}$$

the time taken is the Half Life:

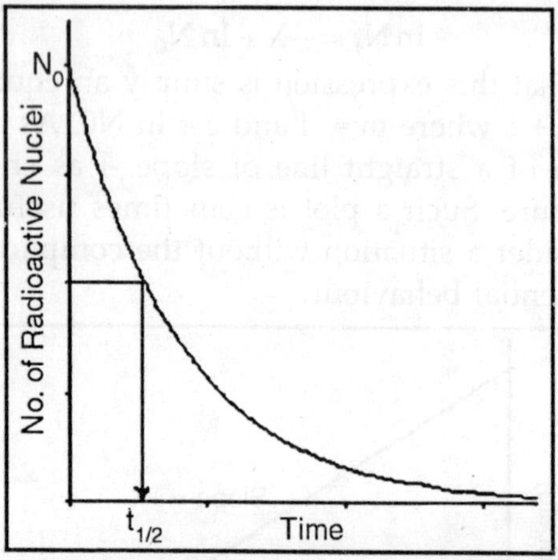

Note that the half-life does not express how long a material will remain radioactive but simply the length of time for its radioactivity to halve. Examples of the half lives of some radioisotopes are given in the following table. Notice that some of these have a relatively short half life. These tend to be the ones used for medical diagnostic purposes because they do not remain radioactive for very long following administration to a patient and hence result in a relatively low radiation dose.

Radioisotope	Half Life (approx.)
^{81m}Kr	13 seconds
ǀTc	6 hours
^{131}I	8 days
^{51}Cr	1 month
^{137}Cs	30 years
^{241}Am	462 years
^{226}Ra	1620 years
^{238}U	4.51×10^9 years

But they do present a logistical problem when we wish to use them when there may not be a radioisotope production facility nearby. For example suppose we wish to use ^{99m}Tc for a patient study and the nearest nuclear facility for making this isotope is 5,000 km away. The production facility could be in Sydney and the patient could be in Perth for instance.

After making the isotope at the nuclear plant it would be decaying with a half life of 6 hours.

So we put the material on a truck and drive it to Sydney airport. The isotope would be decaying as the truck sits in Sydney traffic then decaying still more as it waits for a plane to take it to Perth. Then decaying more as it is flown across to Perth and so on. By the time it gets to our patient it will have substantially reduced in radioactivity possibly to the point of being useless for the patient's investigation. And what about the problem if we needed to use ^{81m}Kr instead of ^{99m}Tc for our patient?

You can appreciate from the table above that other isotopes have a very long half lives. For example^{226}Ra has a half life of over 1,500 years. This isotope has been used in the past for therapeutic applications in medicine. Think about the logistical problems here. They obviously do not relate to transporting the material from the point of production to the point of use.

But they relate to how the material is kept following its arrival at the point of use. We must have a storage facility so that the material can be kept safely for a long period of time. But for how long? A general rule of thumb for the quantities of radioactivity used in medicine is that the radioactivity will remain significant for about 10 half lives. So we would have to have a safe environment for storage of the ^{226}Ra for about 16,000 years! This storage facility would have to be secure from many unforeseeable events such as earthquakes, bombing etc and be kept in a manner which our children's, children's children can understand. A very serious undertaking indeed!

Relationship between Decay Constant/Half Life

On the basis of the above you should be able to appreciate that there is a relationship between the Decay Constant and the Half Life. For example when the Decay Constant is small the Half Life should be long and correspondingly when the Decay Constant is large the Half Life should be short. But what exactly is the nature of this relationship?

We can easily answer this question by using the definition of Half Life and applying it to the Radioactive Decay Law.

Once again the law tells us that at any time, t:

$$N_t = N_0 \exp(-\lambda t)$$

and the definition of Half Life tells us that:

$$N_t = \frac{N_0}{2}$$

when:

$$t = t_{\frac{1}{2}}$$

We can therefore re-write the Radioactive Decay Law by substituting for Nt and t as follows:

$$\frac{N_0}{2} = N_0 \exp(-\lambda t_{\frac{1}{2}})$$

Therefore:

$$\frac{1}{2} = \exp(-\lambda t_{\frac{1}{2}})$$

$$\therefore 2^{-1} = \exp(-\lambda t_{\frac{1}{2}})$$

$$\therefore \text{In} 2^{-1} = -\lambda t_{\frac{1}{2}}$$

$$\therefore \text{In} 2 = \lambda t_{\frac{1}{2}}$$

$$\therefore 0.693 = \lambda t_{\frac{1}{2}}$$

$$t_{\frac{1}{2}} = \frac{0.693}{\lambda}$$

and

$$\lambda = \frac{0.693}{t_{\frac{1}{2}}}$$

These last two equations express the relationship between the Decay Constant and the Half Life. They are very useful as you will see when solving numerical questions relating to radioactivity and usually form the first step in solving a numerical problem.

UNITS OF RADIOACTIVITY

The SI or metric unit of radioactivity is named after Henri Becquerel, in honour of his discovery of radioactivity, and is called the becquerel with the symbol Bq. The becquerel is defined as the quantity of radioactive substance that gives rise to a decay rate of 1 decay per second.

In medical diagnostic work 1 Bq is a rather small amount of radioactivity. Indeed it is easy to remember its definition if you think of it as a buggerall amount of radioactivity. For this reason the kilobecquerel (kBq) and megabecquerel (MBq) are more frequently used.

The traditional unit of radioactivity is named after Marie Curie and is called the curie, with the symbol Ci. The curie is defined as the amount of radioactive substance which gives rise to a decay rate of 3.7×10^{10} decays per second. In other words 37 thousand, million decays per second which as you might appreciate is a substantial amount of radioactivity. For medical diagnostic work the millicurie (mCi) and the microcurie (μCi) are therefore more frequently used.

Why two units? It in essence like all other units of measurement depends on what part of the world you are in. For example the kilometre is widely used in Europe and Australia as a unit of distance and the mile is used in the USA. So if you are reading an American textbook you are likely to find the curie used as the unit of radioactivity, if you are reading an Australian book it will most likely refer to becquerels and both units might be used if you are reading a European book. You will therefore find it necessary to know and understand both units.

Questions

Three questions are given below to help you develop your understanding of the material presented in this chapter. The first one is relatively straight-forward and will exercise your application of the Radioactive Decay Law as well as your understanding of the concept of Half Life. The second question is a lot more challenging and will help you relate the Radioactive Decay Law to the number of radioactive nuclei which are decaying in a sample of radioactive material. The third question will help you understand the approach used in the second question by asking a similar question from a slightly different perspective.

Question **1**

(a) The half-life of ^{99m}Tc is 6 hours. After how much time will 1/16th of the radioisotope remain?
(b) Verify your answer by another means.

Answer

(a) Starting with the relationship we established earlier between the Decay Constant and the Half Life we can calculate the Decay Constant as follows:

$$\lambda = \frac{0.693}{t_{\frac{1}{2}}} = \frac{0.693}{6} = 0.1155\,\text{hr}^{-1}$$

Now applying the Radioactive Decay Law,

$$N_t = N_0 \exp(-\lambda t)$$

we can re-write it in the form:

$$\frac{N_t}{N_0} = \exp(-\lambda t)$$

The question tells us that N_0 has reduced to 1/16th of its value, that is:

$$\frac{N_t}{N_0} = \frac{1}{16}$$

Therefore:

$$\frac{1}{16} = \exp(-0.1155t)$$

which we need to solve for t. One way of doing this is as follows:

$$16^{-1} = \exp(-0.1155t)$$

$$\therefore -\ln 16 = -0.1155t$$

$$t = \frac{\ln 16}{0.1155} = 24\,\text{hours}$$

So it will take 24 hours until 1/16th of the radioactivity remains.

(b) A way in which this answer can be verified is by using the definition of Half Life. We are told that the Half Life of 99mTc is 6 hours. Therefore after six hours half of the radioactivity remains.

Therefore after 12 hours a quarter remains; after 18 hours an eighth remains and after 24 hours one sixteenth remains. And we arrive at the same answer as in part (a).

Note that this second approach is useful if we are dealing with relatively simple situations where the radioactivity is halved, quartered and so on.

But supposing the question asked how long would it take for the radioactivity to decrease to a tenth of its initial value. Deduction from the definition of half life is rather more difficult in this case and the mathematical approach used for part (a) above will yield the answer more readily.

Question 2

Find the radioactivity of a 1 g sample of 226Ra given that t1/2: 1620 years and Avogadro's Number: 6.023 × 1023.

Answer

We can start the answer like we did with Question 1(a) by calculating the Decay Constant from the Half Life using the following equation:

$$\lambda=\frac{0.693}{t_{\frac{1}{2}}}=\frac{0.693}{1620}=4.28\times10^{-4}\ \text{year}^{-1}$$

$$\therefore \lambda=1.36\times10^{-11}\ \text{s}^{-1}$$

Note that the length of a year used in converting from'per year' to'per second' above is 365.25 days to account for leap years.

In addition the reason for converting to units of'per second' is because the unit of radioactivity is expressed as the number of nuclei decaying per second.

Secondly we can calculate that 1 g of ^{226}Ra contains:

$$N=\frac{(\text{Avogadro's No.})(\text{Mass})}{\text{Mass Number}}=\frac{\left(6.023\times10^{23}\right)(1\text{g})}{226}=2.7\times10^{21}\ \text{nuclei}$$

Thirdly we need to express the Radioactive Decay Law in terms of the number of nuclei decaying per unit time.

We can do this by differentiating the equation as follows:

$$N = N_0 \exp(-\lambda t)$$

$$\therefore \frac{dN}{dt} = N_0 . -\lambda \exp(-\lambda t) = -\lambda N_0 \exp(-\lambda t)$$

$$\therefore \frac{dN}{dt} = -\lambda N$$

$$\therefore \left|\frac{dN}{dt}\right| = \lambda N$$

The reason for expressing the result above in absolute terms is to remove the minus sign in that we already know that the number is decreasing.

We can now enter the data we derived above for λ and N:

$$\left|\frac{dN}{dt}\right| = \left(1.36 \times 10^{-11}\right)\left(2.7 \times 10^{21}\right)$$

$$\therefore \left|\frac{dN}{dt}\right| = 3.6 \times 10^{10} \text{ decays per second}$$

So the radioactivity of our 1 g sample of radium-226 is approximately 1 Ci. This is not a surprising answer since the definition of the curie was originally conceived as the radioactivity of 1 g of radium-226!

Question 3

What is the minimum mass of ^{99m}Tc that can have a radioactivity of 1 MBq? Assume the half-life is 6 hours and that Avogadro's Number is 6.023×10^{23}.

Answer

Starting again with the relationship between the Decay Constant and the Half Life.

$$\lambda = \frac{0.693}{6} = 0.1155 \text{ hour}^{-1} = 3.21 \times 10^{-5} \text{s}^{-1}$$

Secondly the question tells us that the radioactivity is 1 MBq. Therefore since 1 MBq = 1×10^6 decays per second,

$$\left|\frac{dN}{dt}\right| = \lambda N = 1 \times 10^6 \text{ dps}$$

$$\therefore N = \frac{\left|\frac{dN}{dt}\right|}{\lambda} = \frac{1\times10^{6}}{3.21\times10^{-5}} = 3.116\times10^{10} \text{ nuclei}$$

Finally the mass of these nuclei can be calculated as follows:

$$\text{Mass of N nuclei} = \frac{(\text{No. of Nuclei})(\text{Mass No.})}{\text{Avogadro's Number}}$$

$$= \frac{(3.116\times10^{10})(99)}{6.023\times10^{23}} = 5.122\times10^{-12} \text{ g}$$

In other words a mass of just over five picograms of ^{99m}Tc can emit one million gamma-rays per second. The result reinforces an important point that you will learn about radiation protection which is that you should treat radioactive materials just like you would handle pathogenic bacteria!

Chapter 6

Radioactive Decay and Exponential Laws

RADIOACTIVE DECAY

In the previous article, we saw that light attenuation obeys an exponential law. To show this, we needed to make one critical assumption: that for a thin enough slice of matter, the proportion of light getting through the slice was proportional to the thickness of the slice.

Exactly the same treatment can be applied to radioactive decay. However, now the"thin slice" is an interval of time, and the dependent variable is the number of radioactive atoms present, N(t).

Radioactive atoms decay randomly. If we have a sample of atoms, and we consider a time interval short enough that the population of atoms hasn't changed significantly through decay, then the proportion of atoms decaying in our short time interval will be proportional to the length of the interval. We end up with a solution known as the"Law of Radioactive Decay", which mathematically is merely the same solution that we saw in the case of light attenuation. We get an expression for the number of atoms remaining, N, as a proportion of the number of atoms N_0 at time 0, in terms of time, t:

$$N/N0 = e^{-lt},$$

where the quantity l, known as the"radioactive decay constant", depends on the particular radioactive substance.

Again, we find a"chance" process being described by an

exponential decay law. We can easily find an expression for the chance that a radioactive atom will"survive" (be an original element atom) to at least a time t. The steps are the same as in the case of photon survival.

MEAN LIFETIME OF A RADIOACTIVE ATOM

On average, how much time will pass before a radioactive atom decays?

This question can be answered using a little bit of calculus. Suppose that we invert our function for N/N_0 in terms of t, to get an expression for t as a function of N/N_0. Once we have an expression for t, a"definite integral" will give us the mean value of t (this is how"mean value" is defined).

From the equation above, taking logarithms of both sides we see that $lt = -\ln(N/N_0) = \ln(N_0/N)$, so our equation for t is,

$$t = t\left(\frac{N}{N_0}\right) = \frac{1}{L}\ln\left(\frac{N_0}{N}\right).$$

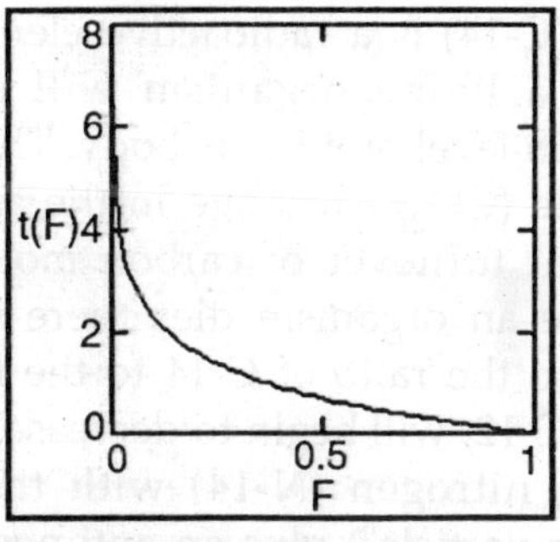

t plotted against F

For convenience, we'll now write F forN/N_0. Note that that the domain of F is the interval from zero to 1, which corresponds to the interval of time from zero to infinity. Plotting t against F with a value of l=1 gives the graph on the right. To find <t>, the mean value of time of survival, all we have to do is find the integral,

$$<t> = \frac{1}{1-0}\int_0^1 \frac{1}{l}\ln\frac{1}{F}\,dF$$

$$= -\frac{1}{l}\int_0^1 \text{In}(F)dF$$

$$= \frac{1}{l},$$

which is a very tidy result!

Incidentally, our formula for t gives us an easy way of finding thehalf-life, the time it takes for half the nuclei in a sample to decay. The half-life (often denoted $t_{1/2}$) is just t(1/2) = (1/l) ln(2). The equivalent thickness for the medium in radiation attenuation is known as"half-value thickness". Similarly, in a population which grows exponentially with time there is the concept of"doubling time".

LIBBY'S LEGACY

We started the first article by talking about carbon dating and the Dead Sea scrolls. Let's look further at the technique behind the work that led to Libby being awarded a Nobel prize in 1960.

Carbon 14 (C-14) is a radioactive element that is found naturally, and a living organism will absorb C-14 and maintain a certain level of it in the body. This is because there is carbon dioxide (CO_2) exchange in the atmosphere, which leads to constant turnover of carbon molecules within the body cells. Once an organism dies there is no further CO_2 exchange, and so the ratio of C-14 to the far more common carbon isotope, C-12, will begin to decrease as the C-14 atoms decay, yielding nitrogen (N-14) with the emission of an electron (or"beta particle") plus an anti-neutrino.

The ratio of C-14 to C-12 in the atmosphere's carbon dioxide molecules is about 1.3×10^{-12}, and this value is assumed constant for the main part of archaeological history since the formation of the earth's atmosphere.

Knowing the level of activity of a sample of organic material enables us to deduce how much C-14 there is in the material at present. Since we also know the ratio of C-14 to C-12 originally, we can find the time that has passed since carbon exchange ceased, that is, since the organic material"died".

In the case of the Dead Sea scrolls, important questions required answers. Were they forgeries? Did they really date from around the time of Christ? Before or after?

Using Libby's radiocarbon dating technique, the scrolls have been dated, using the linen coverings the scrolls were wrapped in. One scroll, the Book of Isaiah, has been dated at 1917BC ±275 years, certainly long before the time of Christ. Some of the others are roughly contemporary with Christ.

Let's take a look at an example of how dates are calculated using Libby's method.

Suppose a linen sample of 1 gram is analysed in a counter. The activity is measured at approximately 11.9 decays per minute. We'll denote the magnitude of the rate of decay of the Carbon 14 nuclei as R. This magnitude is equal to the rate that beta particles are detected. So

$$\frac{dN}{dt} = -R.$$

Recall that the exponential law for the number of Carbon 14 nuclei present says that,

$$N = N_0 / e^{-It}$$

and so
$$\frac{dN}{dt} = -IN_0 / e^{-It} = -IN,$$

which tells us that R=lN, and that the activity at t=0 (the time the linen was manufactured) is $R(0) = lN_0$.

Substituting gives us an exponential relation in terms of the measured activity:

$$R = R_0 e^{It}.$$

Now the decay constant for Carbon-14 is l = 3.8394 × 10-
12 per second. This corresponds to a half life of 5,730 years.

We can calculate the number of Carbon-12 nuclei in 1 gram of carbon:

$$\frac{6.02 \times 10^{-12}\ \text{nuclei / mole}}{12.0\ \text{gram / mole}} \times 1\text{gram} = 5.0167 \times 10^{-22}\ \text{nuclei}.$$

Using the (living) ratio of C-14 to C-12, this implies that the original (t = 0) number of Carbon 14 nuclei was,

$$\left(1.3 \times 10^{-12}\right) \times \left(5.0167 \times 10^{22}\right) = 6.5221 \times 10^{10}.$$

Now, rearranging the exponential activity law gives,

$$t = -\frac{\ln(R/R_0)}{I}.$$

R_0 is simply (3.8394×10-[12])(6.5221×10[10]) = 0.2504 decays per second. The measured rate is R(t) = 11.9 decays per minute = 0.1983 decays per second.

The value for t that results is,

$$6.0758 \times 10^{10} \times \frac{1\,\text{year}}{3.15 \times 10^{7}\,\text{seconds}},$$

which is approximately 1929 years. This an approximate age for some of the scrolls.

In a similar way, dating charcoal found at Stonehenge gives ages of approximately 3798 ±275 years, and, when used on some of the oldest archaeological artefacts in the Americas, the technique gives ages of approximately 25,000 years, corresponding to a time of significantly lower sea levels and supporting the theory that the very first humans in the"New World" crossed the Bering Straits by foot from Siberia into Alaska.

Incidentally, other"larger time scale" radio-dating techniques exist, apart from Libby's radiocarbon method. Here isotopes with longer half lives are used, which enables dating of geological formations and rocks. However, the essential ideas are analogous. For example, in lava form, molten lead and Uranium-238 (standard isotope) are constantly mixed in a certain ratio of their natural abundance. Once solidified, the lead is"locked" in place and since the uranium decays to lead, the lead-to-uranium ratio increases with time. In this way, some of the oldest rocks have been measured at approximately 3 billion years.

EMISSIONS

Emission means letting of something,mostly in the air, anything from body gas to carbon mostly in this case i believe that u are asking about the carbon case. well, any factory that manufactures and produces, the company releases the pollutants of the raw materials.

BREATHING EARTH SIMULATION

Welcome to Breathing Earth. This real-time simulation displays the CO2 emissions of every country in the world, as well as their birth and death rates.

Please remember that this real time simulation is just that: a simulation. Although the CO2 emission, birth rate and death rate data used in Breathing Earth comes from reputable sources, data that measures things on such a massive scale can never be 100% accurate. Please note however that the CO2 emission levels shown here are much more likely to be too low than they are to be too high.

THE ENVIRONMENT AND CLIMATE CHANGE

Global warming (aka climate change) is probably the most important issue to face our generation, and quite possibly any generation in history. The worldwide scientific community is virtually unanimous in its agreement that global warming is happening, that that it's our fault. If we let it get out of our control, the consequences - which will already begin occuring in most of our lifetimes - will be catastrophic. Just some of the consequences that can be reasonably expected are rising sea levels, more frequent and more severe natural disasters, large-scale food shortages, plagues, massive species extinctions, unprecendented numbers of refugees, intensified ethnic and political tensions, and a global economic depression the likes of which no one has ever seen.

The situation is still within our grasp, but we must act now, we must act strongly, and we must act together. Individuals, companies, and governments across the globe must each do what they can to reverse climate change. We will never get a second chance.

The good news is that there are plenty of things that we can do to reduce our carbon footprint. The key word is reduce. We can greatly lessen our impact on climate change by using the planet's resources more responsibly. There are many things we can reduce, and many ways we can reduce them, but three of the major ones are: reduce the amount of animal products you consume (meat, dairy, eggs, leather, etc.), reduce the amount of fuel you use (car, air travel, etc.), and reduce the amount of electricity you use. If you're interested, there are plenty of good resources on the net. I encourage you to so your own resea. useful.

WHERE DOES THE DATA COME FROM

Population: Data is based on July 2008 estimates from the CIA World Factbook. When Breathing Earth is started, it uses each country's birth and death rates to calculate how much its population has changed since July 2008, and adjusts its population figure accordingly. To calculate the total world population, Breathing Earth adds up the population figures of all countries. It continues adjusting the various population figures as you watch it, each time a person is born or a person dies. CO_2 emission rates: figures from the United Nations Statistics Division. These are the most up-to-date figures as of December 2008. Collating CO_2 emissions data for every country on Earth, representing the same time period, is undoubtedly a massive and very complex task that relies on the availability of many other sets of data. This probably explains why the most recent CO_2 emissions data available.

CO_2 emission rates from two years earlier: When Breathing Earth was first built, it used 2002 figures, also from the United Nations Statistics Division. When you hover your mouse over a country, Breathing Earth compares the 2002 and 2004 figures and indicates whether that country's CO_2 emissions have increased or decreased in that time, using the red or green arrow that appears near the bottom-left.

There was an unavailability of a portion of the data for a few of the tinier countries (eg. Vanuatu, Tuvalu, Lesotho). In such cases, An estimates based on their population, economy, and the data of their relevant neighbours. In all such cases, the figures were so low that even had my estimates been wildly inaccurate, the effect on the simulation would have been negligible.

CO_2 EMISSIONS

Some people ask why Breathing Earth focuses on the CO_2

emissions per country, rather than per capita. After all, wouldn't the per capita rates give a better indication of who is being most wasteful? For example, the citizens of Australia, Kuwait and Luxembourg are among the world's worst polluters, yet their CO_2 emissions aren't very prominent on Breathing Earth because of those countries' relatively low populations.

The fact of the matter, however, is that what is most important is how many CO_2 emissions there are from the perspective of Planet Earth. Although some countries are clearly much worse polluters than others, the problem is ultimately a global one that humans of various nationalities have caused, and that humans of various nationalities must work together to solve.

One thing must surely be obvious though: The problem is largely a Western one. It is the Western countries who are leading the way in CO_2 emissions, and when non-Western countries have high CO_2 emission rates themselves, it's usually because they are adopting Western habits. Since we, the West, have been a leading cause of the CO_2 emissions problem, surely it is we who must step up and be the leaders in the solution.

Chapter 7

Introduction to Rays

RAY TRACING

Ray tracing is a process that involves tracing the path of light rays from the perspective of the beholder. It is one method of creating 3D graphics through the use of light. By making use of the path of light, one can create images with extraordinarily life-like or realistic effects.

To render an image through ray tracing, there is a need for a ray tracing programme which handles the full range of calculations needed to properly'render' or portray each individual pixel or square of a designated scene. This programme allows the user to define what the object or the scene will be without the need for having a real object in place, and it permits the user to manipulate the object or scene.

Its most obvious use of is in the field of entertainment - 3D animation movies, computer games, movie special effects and so on.

THE PRINCIPLES BEHIND RAY TRACING

The starting point to understand ray tracing is by understanding how we'see' objects in a scene. The best analogy is to think of oneself as watching a scene (for example, a tabletop with various objects on it) through a camera lens or viewer, with the viewer broken up into individual pixels or points where each point matches one ray of light that connects the beholder's eye to one point on the object.

For example, a 640 x 480 resolution picture means a grid composed of 640 horizontal pixels and 480 vertical ones. Each pixel corresponds to one'spot' (or point) on the scene or'world' that one is looking at. If one is looking at a plain white surface, what he sees is 307,200 white pixels.

However, if one places a solid, three-dimensional object on the white surface (e.g. a red ball), the'picture' becomes a bit more complex. Depending on the angle from which the light is coming, some of the pixels will be plain white; others will have shades of grey going to black (from the shadow cast by light on the object). Other pixels will have varying shades of red because the shape of the object means that light rays striking it are reflected back in different ways.

If the source of light, moreover, is above the viewer, one can expect an area of non-red where light is reflected back at him (think of the'red eyes' which happens when one shoots a picture of someone who is looking straight at his camera lens).

The picture becomes more complex as more objects are added to the'scene' because light interacts with each object. If the original red ball is painted with shiny, reflective paint, one can expect to see reflections from other objects beside or around it. A transparent ball will have both reflective and refractive properties - it will reflect images of something beside it, but it will also show an object behind it (since it is transparent) but the object behind it is affected by the shape of the object one is viewing it through (refraction).

TYPES OF RAYS FROM THE SUN

It is a well known fact that over exposure to the sun can lead to problems, some of which can be fatal. At the same time being exposed to the sun can do wonders for your mood and it is also essential to maintain your health.

Rays from the sun can be very intense and when the sun is at its highest, these rays can cause a great deal of damage to the skin

One of the rays that is emitted from the sun is known as UVC and this is not a real threat to the skin as protective boundary that protects the Earth, the O-Zone layer, does a good job in absorbing such rays.

Another form of ray from the sun is known as UVB. This

generally affects the outer layer of the skin and can cause the skin to burn, so you can see why protecting your skin is very important.

The main ray that is emitted from the sun in UVA and this is the most powerful in that it is able to penetrate pass the upper layer of the skin and therefore damage the lower of the skin. It is this ray that can cause the skin to become burnt and it is this that can lead to the development of skin cancer.

But wait a moment whilst your skin is at risk, there is another part of the body that it is also at risk from the intense rays of the sun. And here we are referring to the eyes.

Your eyes are very delicate indeed and not much has to happen to cause your eyes to become damaged. So you wear lotion to protect your skin so you need to wear sunglasses to protect your eyes.

Your skin is at risk from the sun throughout the year, even during the winter and there is snow on the ground. This might sound a little crazy but all a sheet of ice will do is reflect the rays of the sun back up and into your body, so you need to be aware of the sun throughout the year and not just during the winter months.

What you need to do is to make sure that you are well protected and you will realise that this is a very simple process. Wear the right clothing and make sure that you are protecting your eyes from the glare of the sun.

Also make sure that you are wearing a good sun block or sun lotion. Do not worry too much about the brand, but you need to look at the SPF and make sure that it is no less than 15. If you are very fair skinned then you need to use products where the SPF is no less than 18.

MASS - DEFECT

Careful measurements have shown that the mass of a particular atom is always slightly less than the sum of the masses of the individual neutrons, protons, and electrons of which the atom consists. The difference between the mass of the atom and the sum of the masses of its parts iscalled the mass defect (Dm). The mass defect can be calculated using Equation. In calculating the mass defect it is important to use the full accuracy of mass measurements because the difference in mass is small compared to the mass of the atom. Rounding off the masses of atoms and particles to three or four significant digits prior to the calculation will result in acalculated mass defect of zero.

$$Dm = [Z(m_p + m_e) + (A-Z)m_n] - m_{atom}$$

where:

Dm=mass defect (amu)m_p=mass of a proton (1.007277 amu)

m_n=mass of a neutron (1.008665 amu)m_e=mass of an electron (0.000548597 amu)

m_{atom}=mass of nuclide $^{A}Z\ X$ (amu)

Z=atomic number (number of protons)

A=mass number (number of nucleons)

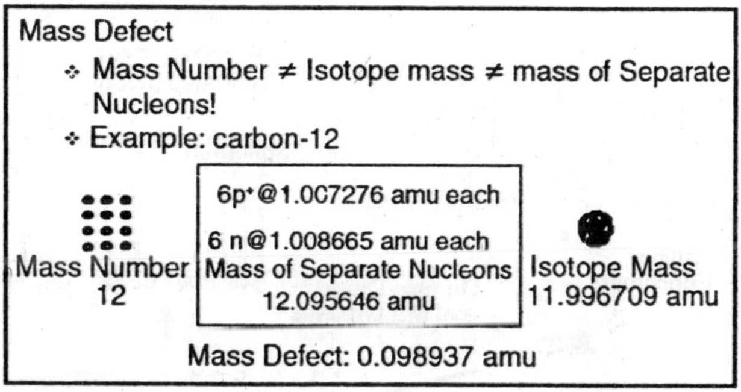

MASS DEFECT TECHNOLOGY

Mass Defect Technology refers to chemical tagging of biomolecules with a chemical compound that contains one or more elements with a substantial mass defect. Because the mass defect causes an element to be lighter in weight that its theoretical nuclear composition suggests, anything tagged with this element is shifted in the mass spectrum, away from untagged biomolecular species (which have near zero mass defects). As such, mass defect technology is broadly applicable to many aspects of biomolecular mass spectrometry and can be incorporated into a variety of mass spectrometric methods.

ISOTOPES

IDBEST allows tagged immunoaffinity-enriched biomarkers to be discriminated in the mass spectrometer from untagged antibodies or bait proteins used for the enrichment. In addition to the mass defect element, IDBEST tags also incorporate stable isotopes to create two chemically-identical, but isotopically-distinct versions of the same tag.

This allows us to tag two patient samples with isotopically-distinct versions of the same tag and mix the tagged patient samples together before the affinity enrichment of the biomarker of interest.

The resulting ratio between the amounts of light (healthy control sample} and heavy (diseased sample) is quantitatively determined as are any differences between peptides containing post-translational modifications.

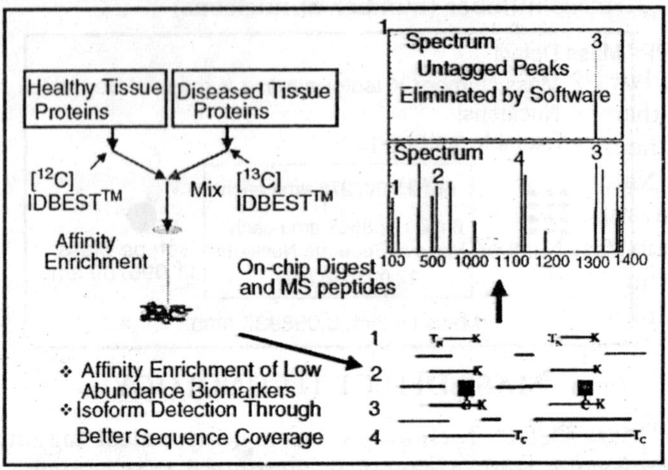

The stable isotope pairing makes the assay quantitative. The mass defect element in the tag allows the biomarker of interest to be discriminated from other extraneous noise in the spectrum using TDI's proprietary deconvolution software. IDBEST™ solves the sample complexity and dynamic range problems intrinsic to global proteomic methods.

- Low abundance biomarkers are selectively enriched and can be seen in the mass spectrometer.
- The sample entering the mass spectrometer is

simplified so that the whole protein can be interrogated for isoforms.

- Multiple biomarker"fishing expeditions" can be conducted on the same sample (i.e., the method is nondestructive of the sample).
- By using affinity enrichment for the protein of interest, the time and cost of analysis is greatly reduced.

MASS DEFECT, FISSION AND FUSION

Mass Defect

When you learned how to calculate the average atomic mass of an element from the exact mass and relative abundance of its isotopes you may have noticed something interesting. The mass number is not the exact mass of an isotope. For example, the exact mass of 16 $_8$O is not exactly 16 u but rather 15.994915 u.The mass of a nucleus is slightly less than you might expect from the total number of protons and neutrons. Why should this be so?

Natural systems follow a consistent pattern in that they always seek to find the lowest possible energy. Atomic nuclei are no exception. As it happens, the nucleus of an atom is thermodynamically more stable than the protons and neutrons that make it up. This can be shown by a relatively simple calculation using the hypothetical situation of making a 16 $_8$O nucleus from 8 hydrogen nuclei and 8 neutrons.

$8^1\ {}_0n^0 + 8^1\ {}_1H \rightarrow {}^{16}{}_8o$

$8n^0 + 8^1\ {}^1H = 8(1.00867) + 8(1.00782) = 16.13192$ u

Actual Mass of ^{16}o = 15.99492 u

Difference in mass: 16.13192 - 15.99492 = 0.13700 u

So an ^{16}o nucleus loses 0.13700 g of mass per mole of ^{16}o formed.

We ignore the electrons because there are as many electrons in 8 H atoms as in one O atom

Table. Exact Masses

$1\ _1H$	1.00782 u
	1 p+ & 1 e-
n^0	1.00867 u
e-	5.48580×10^{-4} u

The mass difference calculated above is called the mass defect. Where does this mass go? It is released as energy when the nucleus forms. The process of forming a nucleus is exothermic: the product is more stable than the starting materials. In other words, the formation of an atomic nucleus releases energy. Energy is in fact a form of matter, as Albert Einstein discovered in the development of his theory of relativity.

You have all seen the famous equation:

$$E = mc^2$$

In this equation a statement about the equivalence of matter and energy is made. It says that energy changes are accompanied by a change in mass equal to the amount of energy divided by c2. In ordinary chemical reactions the amount of energy is so small that the change in mass is not detectable. The c in the equation is a universal constant of nature: the speed of light, 3.00 × 108 m/s. The amount of energy represented by the mass defect is called the binding energy of a nucleus and is the amount of energy that would be required to break the nucleus into its component parts.

Here is how to caculate the amount of energy:

$E = mc^2$

$E = (0.13700\ g/mol \cdot 1\ kg/1000\ g)(3.00 \times 10^8\ m/s)2 = 1.23 \times 10^{13}\ J/mol$

In such calculations grams must be converted to kilograms since 1 joule of energy (1 J) is defined as

1 kg· m^2/s^2. A joule is a scientific unit of energy equal to 4.184×10^{-3} dietary Calories. It is related to the watt (a measure of electrical power) because 1 watt of power uses 1 J per second. A 100-watt light bulb uses 100 J per second. From these considerations and from the above calculation you can see that the binding energy is truly enormous! Perhaps you

now understand why nuclear energy is so interesting: nearly unlimited power at the centre of every atom.

FISSION AND FUSION

Fission is the name for the process in which heavy nuclei (such as$235_{92}U$) split into two nuclei with smaller mass numbers. Fusion is the combination of two light nuclei to form a heavier, more stable nucleus. Both of these processes are highly exothermic and involve the release of energy more than a million times larger than the energy released in chemical reactions.

Fission was discovered in the 1930s neutron bombardment of uranium-235 atoms resulted in the formation of smaller nuclides:

$$1\ {}_0n^0 + {}^{235}{}_{92}U \rightarrow {}^{141}{}_{56}Ba + {}^{92}{}_{36}Kr + 3^1\ {}_0n^0$$

The amount of energy released per mole of uranium-235 is 2.1×10^{13} J. When methane (natural gas) is burned it releases only 8.0×10^5 J/mol. This is a factor of about 26 million times less energy.

Notice that three additional neutrons are produced in the example fission reaction given for uranium-235 (there are actually at least 200 different possible isotopes that can result from such a fission). These neutrons make anuclear chain reaction possible. In a nuclear chain reaction one fission event causes one or more further fission events. When less than one neutron, on average, causes another fission event then the reaction is called sub-critical.

If exactly one neutron from each reaction causes another fission event then the process is self-sustaining and is called critical. Finally, the situation becomes supercritical if more than one neutron from each fission event causes another fission event. This results in a violent explosion, as is well known: a bomb made using a supercritical mass of uranium-235 was dropped on Hiroshima, Japan in 1945. Such bombs are made by arranging for subcritical masses of uranium-235 to be brought together suddenly to form a supercritical mass. Peaceful use of nuclear chain reactions is made in nuclear power generators in which controlled fission chain reactions are used to produce heat, boil water and run steam power generators.

Fusion is the source of the power of the stars. Our Sun is made of hydrogen (73% by mass), helium (26%) and other elements (1%). In the core of the Sun hydrogen nuclei fuse together to form helium nuclei. In the process they release enormous amounts of energy. Seven hundred million tons of hydrogen is fused every second in the core of the Sun and nearly 5 million tons of this matter is released as energy. (Use $E = mc^2$ to figure out how much energy this is; one metric ton is 1,000 kg). The core of the Sun is 16,000,000 degrees Celsius. This high temperature makes earthly power sources based on nuclear fusion problematic at best.

Problems:

- The Sun radiates 3.9×10^{23} J into space every second. What is the rate at which mass is lost from the Sun?
- The earth receives 1.8×10^{17} J/s of solar energy. What mass of solar material is converted to energy in one day to provide this energy to earth (24 hr)? What mass of coal would have to be burned to give the same amount of energy?
- The most stable nucleus in terms of binding energy is ^{56}Fe. If the exact mass of ^{56}Fe is 55.9349 u calculate the binding energy of ^{56}Fe in J/mol.
- Calculate the binding energy in J/mol for carbon-12 (12.00000 u) and uranium-235 (235.0439 u).
- The mass defect for a lithium-6 nucleus is 0.03434 g/mol. Calculate the exact atomic mass of ^6Li.
- The binding energy for a single atom of magnesium-27 is 3.5802×10^{-11} J. Calculate the atomic mass of ^{27}Mg
- The easiest fusion reaction to initiate is,
$$2\ _1H + ^3{}_1H \rightarrow ^4{}_2He + ^1{}_0n_0$$
- Calculate the energy released per ^4He nucleus and per mole of ^4He nuclei. The atomic masses are: ^2H = 2.01410, ^3H = 3.01605, and ^4He = 4.00260. To solve this problem figure out the mass defect of the product nuclei versus the starting nuclei.

- What mass of uranium-235 is required to produce the same amount of energy as a metric ton (1,000 kg) of coal? Assume that the controlled fission of uranium provides about 12.5% of the binding energy of the uranium-235 nuclei. One ton of coal provides about 3.2×10^{10} J.

MASS ENERGY EQUIVALENCE

In this little document we justify Einstein's famous equation $E = mc^2$. Sections 1 and 2 demonstrate how the relativistic mass increase arises. They are essentially non-mathematical and only require the following background information:

- According to the Special Theory of Relativity velocities do not add in the simple way that our"common sense" says.
- The Principle of Inertia, also known as Conservation of Momentum, is true for all collisions.

Sections 3 and 4 extend the discussion. They use only descriptive mathematics, but are non-trivial.

Required background is knowing that:

- In the worldview of Special Relativity spacetime has four dimensions.
- The time squared minus the distance squared is the same number for all observers.

Finally, Section 5 revisits the sticky collisions of the Sections 1 and 2.

NON-RELATIVISTIC STICKY COLLISIONS

In this section we discuss non-relativistic collisions and the Principle of Inertia. Since the discussion is non-relativistic the results should conform to our"common sense" about what should happen.

To the right we show two objects approaching each other at 60 km/hr relative to us. We shall assume that the two objects have equal masses and are constructed of a"sticky" material.

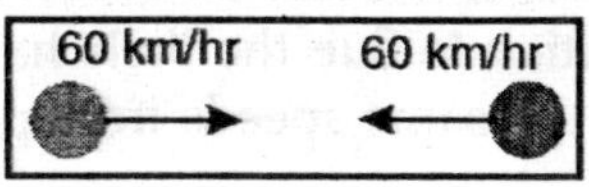

When they two objects collide they stick together. Since before the collision the two objects had equal masses and were moving in opposite directions at the same speeds, their inertia or momenta were equal and opposite. Thus after the collision they will be stationary relative to us. This is an example of the Principle of Inertia, also called Conservation of Momentum.

Now we imagine looking at the same collision for the reference frame of an observer moving to the left at 60 km/hr relative to the first one. The right hand (red) ball will be stationary for this person, but the left hand (blue) ball will be moving at 60 + 60 = 120 km/hr.

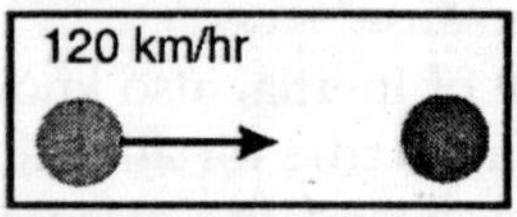

After the collision, the two balls stick together. Since the two balls that are stuck together have twice the mass of the left-hand (blue) ball by itself, after the collision the speed of the two must be one-half of 120 km/hr, i.e. 60 km/hr. Since for the first observer the two balls are stationary after the collision and the second observer is moving to the left at 60 km/hr, we can predict that after the collision the two balls will be moving at 60 km/hr relative to the second observer. So we have two ways of getting the result shown to the right: from Conservation of Momentum or from Galilean relativity.

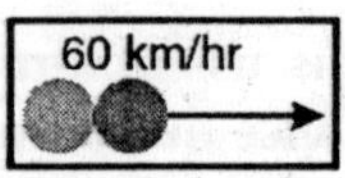

RELATIVISTIC STICKY COLLISIONS

Now we do a similar analysis, but this time the balls are moving at relativistic speeds. Their speeds are 60% of the speed of light relative to the shown observer Sue. We should emphasise that relative to Sue the balls have the same mass and are moving at the same speeds in opposite directions.

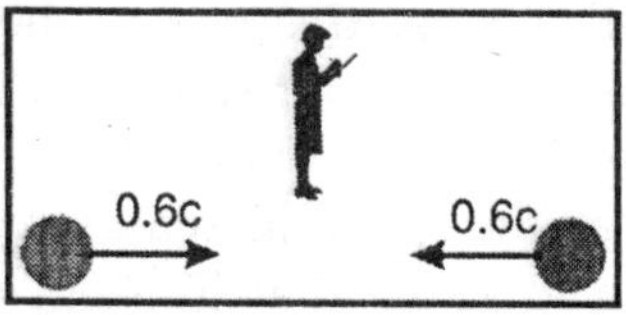

Thus, after the collision the Principle of Inertia says that the balls will be stationary relative to Sue.

Now we imagine an observer Lou who is moving to the left at 60% of the speed of light relative to Sue. The right-hand ball will be stationary relative to Lou. The left-hand ball will be moving to the right relative to him. But because these are relativistic speeds, the left-hand ball will be moving at a speed less than 0.6c + 0.6c = 1.2c. It turns out that the speed of the left-hand ball is 0.88c.

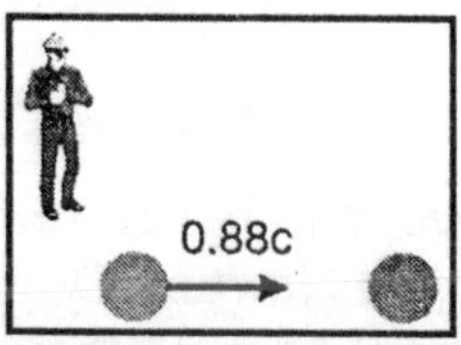

Now we have a problem. The Principle of Inertia predicts that after the collision the speed of the balls will be one-half of 0.88c, i.e. 0.44c. But if the balls were stationary relative to Sue after the collision and Lou is moving to the left at 0.6c relative to Sue, then surely the balls must be moving to the right at 0.6c relative to Lou. So Conservation of Momentum and Galilean relativity predict different results.

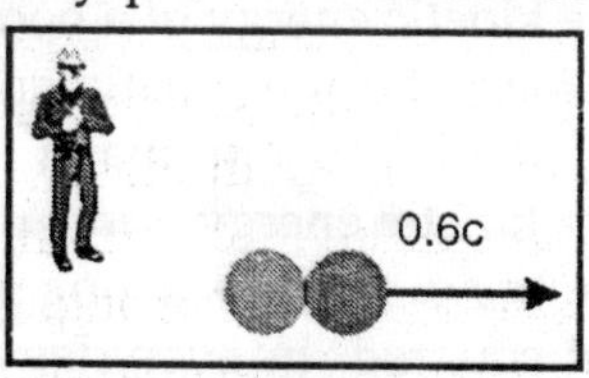

When Einstein was confronted with this, he refused to throw out the Principle of Inertia. Therefore, he concluded that before the collision, relative to Lou the mass of the left-hand ball must be greater than the mass of the right-hand ball. Thus when an object is moving relative to some observer its mass increases as the speed increases.

You may wish to know that Einstein's analysis leading him to the relativistic mass increase was more complex than the above. Instead of considering the interaction of 2 massive objects he was thinking about the interaction between an electromagnetic field and a massive charged particle. This is a more difficult problem but the basic idea is the same.

It turns out that the relation between the mass m the body has when it is moving at speed v relative to an observer is related to the mass m_0 it has relative to an observer for whom the body is at rest is given by the equation shown to the right. The quantity m_0 is called the rest mass of the body. For Lou, before the collision the mass of the left-hand ball is greater than its rest mass by a factor of 2.11.

$$m = \frac{m_0}{\sqrt{1 - v^2 / c^2}}$$

ENERGY

Energy is a concept that has undergone continuous revision since the days of Newton. Originally it referred to a property that bodies have because they are in motion; today that property is called the kinetic energy. Other forms of energy have been discovered since then, with names like potential energy, heat energy, etc. Notice that the kinetic energy is another property, like inertia or momentum, that bodies have because of their motion. Einstein realised that the fact that masses increase when they are moving relative to us means that the kinetic energy of a body could be related to this increase in mass. So the kinetic energy of a body is somehow related to the increase in mass:

$$\text{kinetic energy} \sim m - m_0$$

However, the units of this are wrong. Energy units are a mass times a speed squared. In everyday units, this is a kg times a metre per second squared, which we call a joule. Above we say the energy goes only as the mass.

Relativity taught us that we made a mistake when we measured time in"sacred" units of seconds instead of everyday units of metres. And we learned that we can use the speed of light c as a conversion factor for units. We can apply that realisation to the above to multiply the masses to the speed of light squared to get the units to come out right:

$$\text{kinetic energy} = (m - m_0)\, c^2$$

Einstein showed in 1905 that the above gives the Newtonian form for kinetic energy when the speed of the object v is small compared to the speed of light c. Now, when the object is at rest relative to an observer, it has no kinetic energy.

However what are we to make of the left-over term:

$$m_0\, c^2$$

Einstein realised that this is yet another form of energy, a form that arises because of the body's rest mass. And the total energy of a body that is moving is this"rest energy" plus its kinetic energy.

So:

$$\text{Energy } E = m_0\, c^2 + (m - m_0)\, c^2 = m\, c^2$$

As you probably know, there are other forms of energy than just the rest energy and the kinetic energy. There is heat energy, chemical energy, binding energies of atoms and nuclei, etc. etc. It turns out that all forms of energy are reflected in the total mass of the body. So although we have justified $E = mc^2$ in terms of the kinetic energy, mass-energy equivalence is quite a bit more general.

DIMENSIONALITY OF SPACE-TIME

You will recall that in the worldview of Special Relativity, spacetime has four dimensions, the three spatial dimensions and the time dimension. Here we explore the consequences of this for energy. We shall be slightly mathematical, and will"push the envelope" a bit in terms of sophistication.

Now, the inertia or momentum of a body is related to its mass and its speed. But there is a bit more information than just this. The momentum of an object traveling, say, North is different than the momentum of an object traveling East or an object traveling up. So the momentum has three components, corresponding to the three spatial dimensions. By convention the momentum is given the symbol p, so the three components can be specified as p_x, p_y and p_z

But space-time has four dimensions. What can be the fourth time component of the momentum? A natural candidate is the other quantity that an object has related to its motion, its energy. So it is natural to say the the energy is the fourth component of the momentum.

The dimensions here don't quite work out. Momentum is a mass times a speed, while energy is a mass times a speed squared. But we can once again fix up the units using c. Multiply the spatial momenta by c and they will have the same units as the energy.

So to describe the position of an event occurring in spacetime we specify the four coordinates:

$$\text{Position} = ct, x, y, z$$

To describe the relativistic momentum of an object we similarly specify its four components:

$$\text{Momentum} = E, cp_x, cp_y, cp_z$$

Finally, earlier we saw that although for different observers the time and position of some event may be different, there is a constant for all observers:

$$(ct)^2 - x^2 - y^2 - z^2$$

Similarly, although the momentum and energy of an object is different for different observers, there is a constant here too, the time component squared minus the spatial components squared:

$$E^2 - (cp_x)^2 - (cp_y)^2 - (cp_z)^2$$

What can the value of this quantity be? We could go through all the math, but there is an easier approach. When the object is at rest with respect to us, it has no spatial momentum and its energy is just the rest energy. And if this is the value for an observer at rest relative to the object, then it must be the value for all observers.

$$E^2 - (cp_x)^2 - (cp_y)^2 - (cp_z)^2 = (mc^2)^2$$

In words, the value of the relativistic momentum squared is the rest mass squared.

STICKY COLLISIONS REVISITED

In the sticky collisions, initially the objects are moving and thus have kinetic energy. Relative to the observer for whom the objects are stationary after the collision, the objects have no kinetic energy after the collision.

All the kinetic energy has been converted into heat energy. Relative to other observers, the kinetic energy after the collision is not zero, but is still less than the total kinetic energy before the collision: in this case the difference has also been converted to heat.

But according to $E = mc^2$, energy and mass are equivalent. Thus the heat energy H generated in the collision will have a mass mheat = H/c^2. This is a real mass and has inertia. It turns out that if one tries to apply Conservation of Momentum to the collisions, this additional mass and its momentum must be accounted for in the equations.

A discussion of this plus a recap of Sections 1 and 2 of this document, written for physicists, is available in pdf format here. File size is 66k and it will appear in a separate window.

In 1905, Albert Einstein showed that, as a consequence of his special theory of relativity, mass can be considered to be another form of energy. Thus the law of conservation of energy is really the law of conservation of mass-energy. In normal everyday interactions, the amount of mass that is transferred into other forms of energy (or vice versa) is such a tiny fraction of the total mass that it is beyond our sensory perceptions and measurement techniques.

Thus, in a chemical reaction, for example, mass and energy truly seem to be separately conserved. In a nuclear reaction, however, the energy released is often about a million times greater than in a chemical reaction, and the change in mass can easily be measured.

Mass and energy are related by what is certainly the best-known equation in physics:

$$E=mc^2$$

in which E is the energy equivalent (called mass energy) of mass m, and c is the speed of light.

DERIVATION OF E=MC²

Because energy is the integral of force with respect to distance, kinetic energy K can be represented by the following expression:

$$K = \int_0^S F ds$$

where F is the component of the applied force in the direction of the displacement ds and s is the distance over which the force acts. Using the relativistic form of the second law of motion

$$F = \frac{d(mv)}{dt}$$

the expression for kinetic energy becomes:

$$K = \int_0^S \frac{d(mv)}{dt} ds = \int_0^{mv} v d(mv) = \int_0^v v d\left(\frac{m_0 v}{\sqrt{1 - v^2/c^2}}\right)$$

Integrating by parts:

$$\int x dy = xy - \int y dx$$

yields,

$$K = \frac{m_0 v^2}{\sqrt{1 - v^2/c^2}} - m_0 \int_0^v \frac{v dv}{\sqrt{1 - v^2/c^2}}$$

$$= \frac{m_0 v^2}{\sqrt{1 - v^2/c^2}} + \left[m_0 c^2 \sqrt{1 - v^2/c^2}\right]_0^v$$

$$= \frac{m_0 c^2}{\sqrt{1 - v^2/c^2}} - m_0 c^2$$

$$= mc^2 - m_0 c^2$$

This result states that the kinetic energy of a body is equal to the increase in its mass consequent upon its relative motion multiplied by the square of the speed of light.

This result may also be written as,

$$mc^2 = m_0c^2 + K$$

If we interpret mc2as the total energy E of the body, we see that when the body is at rest and K=0, it still possesses the energy m_0c^2. Accordingly m_0c^2 is called the rest energy $E_0$of a body whose mass at rest is m_0.

We therefore have:

$$E=E_0+k$$

where:

$$E_0=m_0c^2$$

If the body is moving, its total energy is:

$$E = mc^2 = \frac{m_0c^2}{1 - v^2/c^2}$$

NUCLEAR FISSION

NUCLEAR FISSION AND NUCLEAR FUSION

There are two types of atomic explosions that can be facilitated by Uranium-235: fission and fusion. Fission, simply put, is a nuclear reaction in which an atomic nucleus splits into fragments, usually two fragments of comparable mass, emitting 100 million to several hundred million volts of energy. This energy is expelled explosively and violently in the atomic bomb. A fusion reaction is usually started with a fission reaction, but unlike the fission (atomic) bomb, the fusion (hydrogen) bomb derives its power from the fusing of nuclei of various hydrogen isotopes into helium nuclei. This article discusses the A-bomb or atomic bomb.

The massive power behind the reaction in an atomic bomb arises from the forces that hold the atom together. These forces are akin to, but not quite the same as, magnetism.

ABOUT ATOMS

Atoms are comprised of various numbers and combinations of the three sub-atomic particles: protons, neutrons and electrons. Protons and neutrons cluster together to form the nucleus (central mass) of the atom while the electrons orbit the nucleus much like planets around a sun. It is the balance and arrangement of these particles that determine the stability of the atom.

SPLITABILITY

Most elements have very stable atoms which are impossible to split except by bombardment in particle accelerators. For all practical purposes, the only natural element whose atoms can be split easily is uranium, a heavy metal with the largest atom of all natural elements and an unusually high neutron-to-proton ratio. This higher ratio does not enhance its"splitability," but it does have an important bearing on its ability to facilitate an explosion, making uranium-235 an exceptional candidate for nuclear fission.

URANIUM ISOTOPES

There are two naturally-occurring isotopes of uranium. Natural uranium consists mostly of isotope U-238, with 92 protons and 146 neutrons (92+146=238) per atom. Mixed with this is a 0.6% accumulation of U-235, with only 143 neutrons per atom. The atoms of this lighter isotope can be split, thus it is"fissionable" and useful in making atomic bombs. Neutron-heavy U-238 has a role to play in the atomic bomb as well since its neutron-heavy atoms can deflect stray neutrons, preventing an accidental chain reaction in a uranium bomb and keeping neutrons contained in a plutonium bomb. U-238 can also be"saturated" to produce plutonium (Pu-239), a man-made, radioactive element also used in atomic bombs.

Both isotopes of uranium are naturally radioactive; their bulky atoms disintegrating over time. Given enough time (hundreds of thousands of years) uranium will eventually lose so many particles that it will turn into lead. This process of decay can be greatly accelerated in what is known as a chain reaction. Instead of disintegrating naturally and slowly, the atoms are forcibly split by bombardment with neutrons.

CHAIN REACTIONS

A blow from a single neutron is enough to split the less-stable U-235 atom, creating atoms of smaller elements (often barium and krypton) and releasing heat and gamma radiation (the most powerful and lethal form of radioactivity). The chain reaction occurs when"spare" neutrons from this atom fly out with sufficient force to split other U-235 atoms they come in contact with. In theory, it is necessary to split only one U-235 atom, which will release neutrons which will split other atoms, which will release neutrons... and so on. This progression is not arithmetic; it is geometric and takes place within a millionth of a second.

The minimum amount to start a chain reaction as described above is known as super critical mass. For pure U-235, it is 110 pounds (50 kilograms). No uranium is ever quite pure, however, so in reality more will be needed. U-235, U-238 and Plutonium.

ABOUT PLUTONIUM

Uranium is not the only material used for making atomic bombs. Another material is the Pu-239 isotope of the man-made element plutonium. Plutonium is only found naturally in minute traces, so useable amounts must be produced from uranium. In a nuclear reactor, uranium's heavier U-238 isotope can be forced to acquire extra particles, eventually becoming the plutonium.

Plutonium will not start a fast chain reaction by itself, but this difficulty is overcome by having a neutron source, a highly radioactive material that gives off neutrons faster than the Plutonium itself. In certain types of bombs, a mixture of the elements Beryllium and Polonium is used to bring about this reaction. Only a small piece is needed (super critical mass is about 32 pounds, though as little as 22 can be used). The material is not fissionable in and of itself, but merely acts as a catalyst to the greater reaction.

Chapter 8

Nuclear Reactors

OVERVIEW

Researchers at NASA and the Department of Energy recently tested key technologies for developing a nuclear fission reactor that could power a human outpost on the moon or Mars. The tests prove that the agencies could build a"safe, reliable, and efficient" system by 2020, the year NASA plans to return humans to the moon.

A fission reactor works by splitting atoms and releasing energy in the form of heat, which is converted into electricity. The idea for using nuclear power in space dates back to the late 1950s, when they were considered for providing propulsion through Project Orion. In the 1960s a series of compact, experimental space nuclear reactors were developed by NASA under the Systems Nuclear Auxiliary Power programme. But public safety concerns and an international treaty banning nuclear weapons in space stopped development.

Now nuclear power is being considered for lunar and Mars missions because, unlike alternatives such as solar power, it can provide constant energy, a necessity for human life-support systems, recharging rovers, and mining for resources.

Solar power systems would also require the use of energy storage devices like batteries or fuel cells, adding unwanted mass to the system. Solar power is further limited because the moon is dark for up to 14 days at a time and has deep craters that can obscure the sun. Mars is farther away from the sun than either the Earth or the moon, so less solar power can be harvested there.

The new nuclear power system is part of a NASA project started in 2006, called Fission Surface Power, that is examining small reactors designed for use on other planets. While nuclear power remains controversial, the researchers say that the reactor would be designed to be completely safe and would be buried a safe distance from the astronauts to shield them from any radiation it would generate.

The recent tests examined technologies that would see a nuclear reactor coupled with a Stirling engine capable of producing 40 kilowatts of energy--enough to power a future lunar or Mars outpost.

"We are not building a system that needs hundreds of gigawatts of power like those that produce electricity for our cities," says Don Palac, the project manager at NASA Glenn Research Centre in Cleveland, OH. The system needs to be cheap, safe, and robust and"our recent tests demonstrated that we can successfully build that," says Palac. To generate electricity, the researchers used a liquid metal to transfer the heat from the reactor to the Stirling engine, which uses gas pressure to convert heat into the energy needed to generate electricity. For the tests, the researchers used a non-nuclear heat source. The liquid metal was a sodium potassium mixture that has been used in the past to transfer heat from a reactor to a generator, says Palac, but this is the first time this mixture has been used with a Stirling engine.

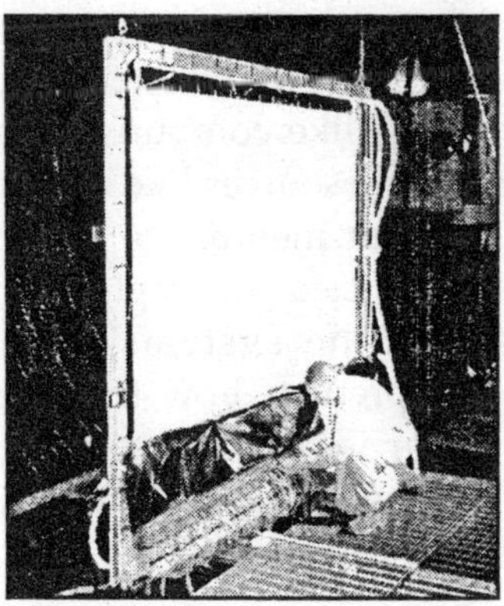

The researchers also developed a lightweight radiator panel to cool the system and dissipate the heat from the reactor. The prototype panel is approximately six feet by nine feet--one-twentieth the size required for a full-scale system. Heat from a water-cooling system is circulated to the radiator where it dissipates.

The researchers tested the radiator panel in a vacuum chamber at Glenn that replicates the lack of atmosphere and the extreme temperatures on the moon--from over 100 degrees Celsius during the day to below 100 degrees Celsius at night. The panel dissipated six kilowatts of energy, more than expected--a"very successfully test," says Palac. On the moon, the panel must also survive the dusty environment cause by the regolith.

Lastly, the researchers tested the performance of the Stirling alternator in a radiation environment at Sandia National Laboratories in Albuquerque, NM. The objective was to test the performance of the motor, ensuring that the materials would not degrade. The alternator was subjected to 20 times the amount of radiation it would expect to see in its lifetime and survived without any significant problems.

Mason says that the tests are very important in showing the feasibility of the system and that the next step is for the researchers to conduct a full system demonstration, by combining a non-nuclear reactor simulator with the Stirling engine and radiator panel. He says that these tests should be completed in 2014.

The researchers are also working on the power transmission and electronics of the system."A lunar base needs lots of power for things like computers, life support, and to heat up rocks to get out resources like oxygen and hydrogen," says Ross Radel, a senior member of the technical staff and part of the advanced nuclear concepts group at Sandia. His group is working on the systems dynamic analysis, a computer model that predicts how the reactor will perform during testing."Nuclear is a stepping stone to move further out into manned space exploration," says Radel.

"It is a fascinating project and the only possible method

of providing power for a manned trip to Mars," says Daniel Hollenbach, a researcher in the nuclear science and technology division at Oak Ridge National Laboratory, who was not involved in the project.

Mason says that nuclear fission is one of a number of concepts being tested as a power source for human missions to the moon and Mars, and if selected, he says the technology could be deployed by 2020.

NUCLEAR POWER REACTORS

- Most nuclear electricity is generated using just two kinds of reactors which were developed in the 1950s and improved since.
- New designs are coming forward and some are in operation as the first generation reactors come to the end of their operating lives.
- Over 16% of the world's electricity is produced from nuclear energy, more than from all sources worldwide in 1960.

A nuclear reactor produces and controls the release of energy from splitting the atoms of certain elements. In a nuclear power reactor, the energy released is used as heat to make steam to generate electricity. (In a research reactor the main purpose is to utilise the actual neutrons produced in the core. In most naval reactors, steam drives a turbine directly for propulsion.)

The principles for using nuclear power to produce electricity are the same for most types of reactor. The energy released from continuous fission of the atoms of the fuel is harnessed as heat in either a gas or water, and is used to produce steam. The steam is used to drive the turbines which produce electricity (as in most fossil fuel plants).

There are several components common to most types of reactors:

FUEL

Usually pellets of uranium oxide (UO2) arranged in tubes to form fuel rods. The rods are arranged into fuel assemblies in the reactor core.*

In a new reactor with new fuel a neutron source is needed to get the reaction going. Usually this is beryllium mixed with polonium, radium or other alpha-emitter. Alpha particles from the decay cause a release of neutrons from the beryllium as it turns to carbon-12. Restarting a reactor with some used fuel may not require this, as there may be enough neutrons to achieve criticality when control rods are removed.

Moderator

This is material in the core which slows down the neutrons released from fission so that they cause more fission. It is usually water, but may be heavy water or graphite.

Control Rods

These are made with neutron-absorbing material such as cadmium, hafnium or boron, and are inserted or withdrawn from the core to control the rate of reaction, or to halt it.* In some reactors, special control rods are used to enable the core to sustain a low level of power efficiently. (Secondary shutdown systems involve adding other neutron absorbers, usually as a fluid, to the system.) In fission, most of the neutrons are released promptly, but some are delayed. These are crucial in enabling a chain reacting system (or reactor) to be controllable and to be able to be held precisely critical.

Coolant

A liquid or gas circulating through the core so as to transfer the heat from it.. In light water reactors the water moderator functions also as primary coolant. Except in BWRs, there is secondary coolant circuit where the steam is made.

Pressure Vessel or Pressure Tubes

Usually a robust steel vessel containing the reactor core and moderator/coolant, but it may be a series of tubes holding the fuel and conveying the coolant through the moderator.

Steam Generator

Part of the cooling system where the primary coolant

bringing heat from the reactor is used to make steam for the turbine.

Containment

The structure around the reactor core which is designed to protect it from outside intrusion and to protect those outside from the effects of radiation in case of any malfunction inside. It is typically a metre-thick concrete and steel structure.

There are several different types of reactors as indicated in the following table.

Table.Nuclear Power Plants in Commercial Operation

Reactor type	Main Countries	Number	GWe	Fuel	Coolant	Moderator
Pressurised Water Reactor (PWR)	US, France, Japan, Russia, China	265	251.6	enriched UO_2	water	water
Boiling Water Reactor (BWR)	US, Japan, Sweden	94	86.4	enriched UO_2	water	water
Pressurised Heavy Water Reactor 'CANDU' (PHWR)	Canada	44	24.3	natural UO_2	heavy water	heavy water
Gas-cooled Reactor (AGR & Magnox)	UK	18	10.8	natural U (metal), enriched UO_2	CO_2	graphite
Light Water Graphite Reactor (RBMK)	Russia	12	12.3	enriched UO_2	water	graphite
Fast Neutron Reactor (FBR)	Japan, France, Russia	4	1.0	PuO_2 and UO_2	liquid sodium	none
Other	Russia	4	0.05	enriched UO_2	water	graphite
	TOTAL	441	386.5			

Most reactors need to be shut down for refuelling, so that the pressure vessel can be opened up. In this case refuelling is at intervals of 1-2 years, when a quarter to a third of the fuel assemblies are replaced with fresh ones. The CANDU and RBMK types have pressure tubes (rather than a pressure vessel enclosing the reactor core) and can be refuelled under load by disconnecting individual pressure tubes.

If graphite or heavy water is used as moderator, it is possible to run a power reactor on natural instead of enriched uranium. Natural uranium has the same elemental composition as when it was mined (0.7% U-235, over 99.2% U-238), enriched uranium has had the proportion of the fissile isotope (U-235) increased by a process called enrichment, commonly to 3.5 - 5.0%. In this case the moderator can be ordinary water, and such reactors are collectively called light water reactors. Because the light water absorbs neutrons as well as slowing them, it is less efficient as a moderator than heavy water or graphite.

Practically all fuel is ceramic uranium oxide (UO_2 with a melting point of 2800°C) and most is enriched. The fuel pellets (usually about 1 cm diameter and 1.5 cm long) are typically arranged in a long zirconium alloy (zircaloy) tube to form a fuel rod, the zirconium being hard, corrosion-resistant and permeable to neutrons.* Numerous rods form a fuel assembly. which is an open lattice and can be lifted into and out of the reactor core. In the most common reactors these are about 3.5 to 4 metres long.

*Zirconium is an important mineral for nuclear power, where it finds its main use. It is therefore subject to controls on trading. It is normally contaminated with hafnium, a neutron absorber, so very pure'nuclear grade' Zr is used to make the zircaloy, which is about 98% Zr plus tin, iron, chromium and sometimes nickel to enhance its strength.

Burnable poisons are often used (especially in BWR) in fuel or coolant to even out the performance of the reactor over time from fresh fuel being loaded to refuelling. These are neutron absorbers which decay under neutron exposure, compensating for the progressive build up of neutron absorbers in the fuel as it is burned. The best known is gadolinium, which is a vital ingredient of fuel in naval reactors where installing fresh fuel is very inconvenient, so reactors are designed to run more than a decade between refuellings.

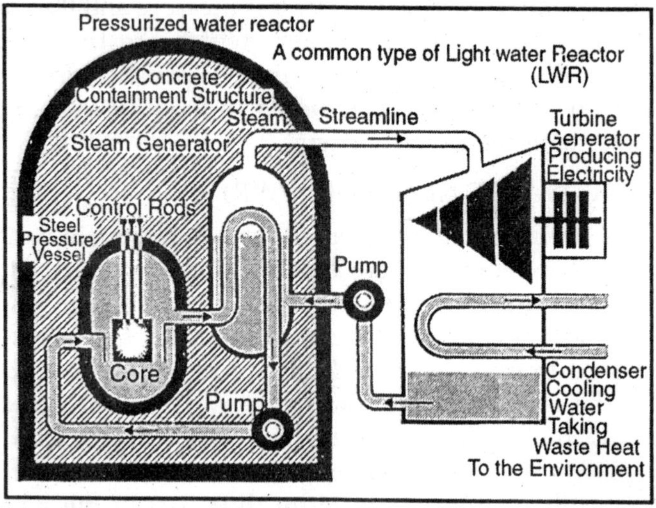

PRESSURISED WATER REACTOR (PWR)

This is the most common type, with over 230 in use for power generation and several hundred more employed for naval propulsion.

The design of PWRs originated as a submarine power plant. PWRs use ordinary water as both coolant and moderator. The design is distinguished by having a primary cooling circuit which flows through the core of the reactor under very high pressure, and a secondary circuit in which steam is generated to drive the turbine.

A PWR has fuel assemblies of 200-300 rods each, arranged vertically in the core, and a large reactor would have about 150-250 fuel assemblies with 80-100 tonnes of uranium.

Water in the reactor core reaches about 325°C, hence it must be kept under about 150 times atmospheric pressure to prevent it boiling. Pressure is maintained by steam in a pressuriser (see diagram). In the primary cooling circuit the water is also the moderator, and if any of it turned to steam the fission reaction would slow down. This negative feedback effect is one of the safety features of the type. The secondary shutdown system involves adding boron to the primary circuit.

The secondary circuit is under less pressure and the water here boils in the heat exchangers which are thus steam generators. The steam drives the turbine to produce electricity, and is then condensed and returned to the heat exchangers in contact with the primary circuit.

BOILING WATER REACTOR (BWR)

This design (diagram next page) has many similarities to the PWR, except that there is only a single circuit in which the water is at lower pressure (about 75 times atmospheric pressure) so that it boils in the core at about 285°C. The reactor is designed to operate with 12-15% of the water in the top part of the core as steam, and hence with less moderating effect and thus efficiency there. The steam passes through drier plates (steam separators) above the core and then directly to the turbines, which are thus part of the reactor circuit. Since the water around the core of a reactor is always contaminated with traces of radionuclides, it means that the turbine must be shielded and radiological protection provided during maintenance. The cost of this tends to balance the savings due to the simpler design. Most of the radioactivity in the water is very short-lived, so the turbine hall can be entered soon after the reactor is shut down.

A BWR fuel assembly comprises 90-100 fuel rods, and there are up to 750 assemblies in a reactor core, holding up to 140 tonnes of uranium. The secondary control system involves restricting water flow through the core so that more steam in the top part reduces moderation.

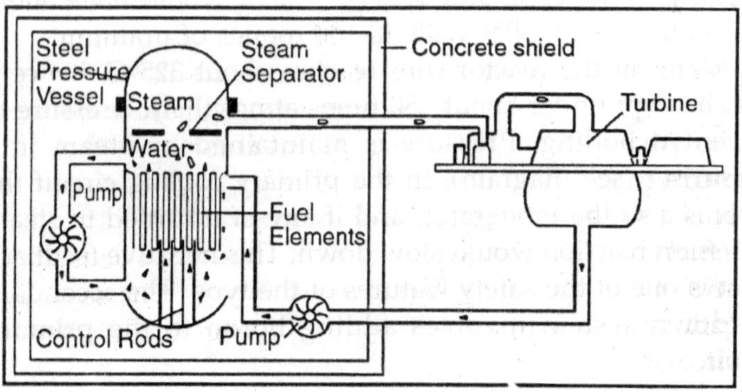

PRESSURISED HEAVY WATER REACTOR

The PHWR reactor design has been developed since the 1950s in Canada as the CANDU, and more recently also in India. It uses natural uranium (0.7% U-235) oxide as fuel, hence needs a more efficient moderator, in this case heavy water (D_2O). with the CANDU system, the moderator is enriched (ie water) rather than the fuel, - a cost trade-off.

The moderator is in a large tank called a calandria, penetrated by several hundred horizontal pressure tubes which form channels for the fuel, cooled by a flow of heavy water under high pressure in the primary cooling circuit, reaching 290°C. As in the PWR, the primary coolant generates steam in a secondary circuit to drive the turbines. The pressure tube design means that the reactor can be refuelled progressively without shutting down, by isolating individual pressure tubes from the cooling circuit.

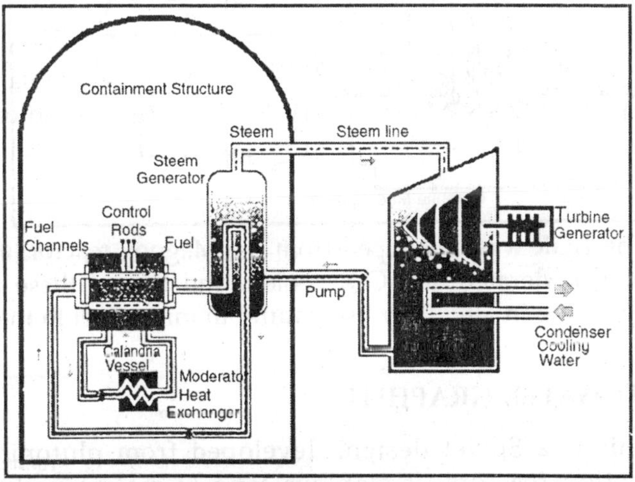

A CANDU fuel assembly consists of a bundle of 37 half metre long fuel rods (ceramic fuel pellets in zircaloy tubes) plus a support structure, with 12 bundles lying end to end in a fuel channel. Control rods penetrate the calandria vertically, and a secondary shutdown system involves adding gadolinium to the moderator. The heavy water moderator circulating through the body of the calandria vessel also yields some heat (though this circuit is not shown on the diagram above).

ADVANCED GAS-COOLED REACTOR (AGR)

These are the second generation of British gas-cooled reactors, using graphite moderator and carbon dioxide as coolant. The fuel is uranium oxide pellets, enriched to 2.5-3.5%, in stainless steel tubes. The carbon dioxide circulates through the core, reaching 650°C and then past steam generator tubes outside it, but still inside the concrete and steel pressure vessel. Control rods penetrate the moderator and a secondary shutdown system involves injecting nitrogen to the coolant.

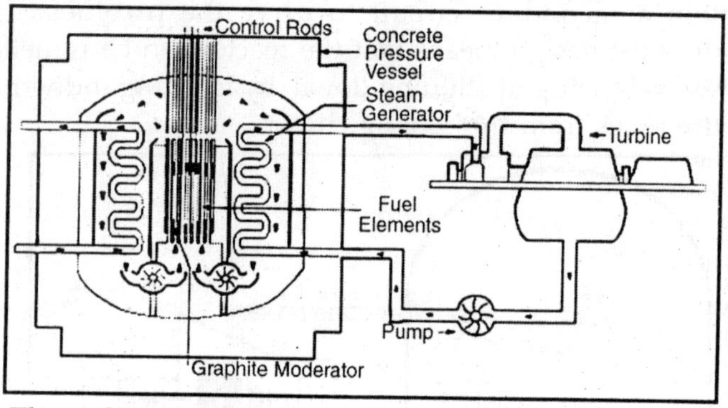

The AGR was developed from the Magnox reactor, also graphite moderated and CO_2 cooled, and two of these are still operating in UK. They use natural uranium fuel in metal form.

LIGHT WATER GRAPHITE

This is a Soviet design, developed from plutonium production reactors. It employs long (7 metre) vertical pressure tubes running through graphite moderator, and is cooled by water, which is allowed to boil in the core at 290°C, much as in a BWR. Fuel is low-enriched uranium oxide made up into fuel assemblies 3.5 metres long. With moderation largely due to the fixed graphite, excess boiling simply reduces the cooling and neutron absorbtion without inhibiting the fission reaction, and a positive feedback problem can arise, which is why they have never been built outside the Soviet Union.

ADVANCED REACTORS

Several generations of reactors are commonly distinguished. Generation I reactors were developed in 1950-60s and very few are still running today. They mostly used natural uranium fuel and used graphite as moderator. Generation II reactors are typified by the present US fleet and most in operation elsewhere. They typically use enriched uranium fuel and are mostly cooled and moderated by water. Generation III are the Advanced Reactors, the first few of which are in operation in Japan and others are under construction and ready to be ordered. They are developments of the second generation with enhanced safety.

Generation IV designs are still on the drawing board and will not be operational before 2020 at the earliest, probably later. They will tend to have closed fuel cycles and burn the long-lived actinides now forming part of spent fuel, so that fission products are the only high-level waste. Many will be fast neutron reactors.

More than a dozen (Generation III) advanced reactor designs are in various stages of development. Some are evolutionary from the PWR, BWR and CANDU designs above, some are more radical departures. The former include the Advanced Boiling Water Reactor, a few of which are now operating with others under construction. The best-known radical new design is the Pebble Bed Modular Reactor, using helium as coolant, at very high temperature, to drive a turbine directly.

Considering the closed fuel cycle, Generation 1-3 reactors recycle plutonium (and possibly uranium), while Generation IV are expected to have full actinide recycle.

FAST NEUTRON REACTORS (FNR)

Some reactors (only one in commercial service) do not have a moderator and utilise fast neutrons, generating power from plutonium while making more of it from the U-238 isotope in or around the fuel. While they get more than 60 times as much energy from the original uranium compared with the normal reactors, they are expensive to build. Further development of them is likely in the next decade, and the main designs expected to be built in two decades are FNRs.

FLOATING NUCLEAR POWER PLANTS

Apart from over 200 nuclear reactors powering various kinds of ships, Rosatom in Russia has set up a subsidiary to supply floating nuclear power plants ranging in size from 70 to 600 MWe. These will be mounted in pairs on a large barge, which will be permanently moored where it is needed to supply power and possibly some desalination to a shore settlement or industrial complex. The first will have two 40 MWe reactors based on those in icebreakers and will operate at Severodvinsk, in the Archangel region. Five of the next seven will be used by Gazprom for offshore oil and gas field development and for operations on the Kola and Yamal peninsulas. One is for Pevek on the Chukotka peninsula, another for Kamchatka region, both in the far east of the country. Further far east sites being considered are Yakutia and Taimyr. Electricity cost is expected to be much lower than from present alternatives.

The Russian KLT-40S is a reactor well proven in icebreakers and now proposed for wider use in desalination and, on barges, for remote area power supply. Here a 150 MWt unit produces 35 MWe (gross) as well as up to 35 MW of heat for desalination or district heating. These are designed to run 3-4 years between refuelling and it is envisaged that they will be operated in pairs to allow for outages, with on-board refuelling capability and used fuel storage. At the end of a 12-year operating cycle the whole plant is taken to a central facility for overhaul and removal of used fuel. Two units will be mounted on a 20,000 tonne barge. A larger Russian factory-built and barge-mounted reactor is the VBER-150, of 350 MW thermal, 110 MWe. The larger VBER-300 PWR is a 325 MWe unit, originally envisaged in pairs as a floating nuclear power plant, displacing 49,000 tonnes. As a cogeneration plant it is rated at 200 MWe and 1900 GJ/hr.

LIFETIME OF NUCLEAR REACTORS

Most of today's nuclear plants which were originally designed for 30 or 40-year operating lives. However, with major investments in systems, structures and components lives can be extended, and in several countries there are active programmes to extend operating lives.

In the USA most of the more than one hundred reactors are expected to be granted licence extensions from 40 to 60 years. This justifies significant capital expenditure in upgrading systems and components, including building in extra performance margins.

Some components simply wear out, corrode or degrade to a low level of efficiency. These need to be replaced. Steam generators are the most prominent and expensive of these, and many have been replaced after about 30 years where the reactor otherwise has the prospect of running for 60 years. This is essentially an economic decision.

Lesser components are more straightforward to replace as they age. In Candu reactors, pressure tube replacement has been undertaken on some plants after about 30 years operation.

A second issue is that of obsolescence. For instance, older reactors have analogue instrument and control systems. Thirdly, the properties of materials may degrade with age, particularly with heat and neutron irradiation. In respect to all these aspects, investment is needed to maintain reliability and safety. Also, periodic safety reviews are undertaken on older plants in line with international safety conventions and principles to ensure that safety margins are maintained.

LOAD-FOLLOWING CAPACITY

Nuclear power plants are essentially base-load generators, running continuously. This is because their power output cannot readily be ramped up and down on a daily and weekly basis, and in this respect they are similar to most coal-fired plants. (It is also uneconomic to run them at less than full capacity, since they are expensive to build but cheap to run.) However, in some situations it is necessary to vary the output according to daily and weekly load cycles on a regular basis, for instance in France, where there is a very high reliance on nuclear power.

The ability of a power reactor to run at less than full power for much of the time depends on whether it is in the early part of its 18 to 24-month refueling cycle or late in it, and whether it is designed with special control rods which diminish power levels throughout the core without shutting it down.

Thus, though the ability on any individual reactor to run on a sustained basis at low power decreases markedly as it progresses through the refueling cycle, there is considerable scope for running a fleet of reactors in load-following mode.

PRIMARY COOLANTS

The advent of some of the designs mentioned above provides opportunity to review the various primary coolants used in nuclear reactors. There is a wide variety - gas, water, light metal, heavy metal and salt:

Water or heavy water must be maintained at very high pressure (1000-2200 psi, 7-15 MPa) to enable it to function above 100°C, as in present reactors. This has a major influence on reactor engineering. However, supercritical water around 25 MPa can give 45% thermal efficiency - as at some fossil-fuel power plants today with outlet temperatures of 600°C, and at ultra supercritical levels (30+ MPa) 50% may be attained.

Helium must be used at similar pressure (1000-2000 psi, 7-14 MPa) to maintain sufficient density for efficient operation. Again, there are engineering implications, but it can be used in the Brayton cycle to drive a turbine directly.

Carbon dioxide was used in early British reactors and their AGRs. It is denser than helium and thus likely to give better thermal conversion efficiency. There is now interest in supercritical CO_2 for the Brayton cycle.

Sodium, as normally used in fast neutron reactors, melts at 98°C and boils at 883°C at atmospheric pressure, so despite the need to keep it dry the engineering required to contain it is relatively modest. However, normally water/steam is used in the secondary circuit to drive a turbine (Rankine cycle) at lower thermal efficiency than the Brayton cycle.

Lead or lead-bismuth eutectic in fast neutron reactors are capable of higher temperature operation. They are transparent to neutrons, aiding efficiency, and since they do not react with water the heat exchanger interface is safer. They do not burn when exposed to air. However, they are corrosive of fuel cladding and steels, and Pb-Bi yields toxic polonium (Po-210) activation products. Pb-Bi melts at a relatively low 125°C (hence eutectic) and boils at 1670°C, Pb melts at 327°C and boils at 1737°C but is very much more abundant and cheaper to produce than bismuth, hence is envisaged for large-scale use in the future. The development of nuclear power based on Pb-Bi cooled fast neutron reactors is likely to be limited to a total of 50-100 GWe, basically for small reactors in remote places. In 1998 Russia declassified a lot of research information derived from its experience with submarine reactors, and US interest in using Pb or Pb-Bi for small reactors has increased subsequently.

Molten fluoride salt boils at 1400°C at atmospheric pressure, so allows several options for use of the heat, including using helium in a secondary Brayton cycle with thermal efficiencies of 48% at 750°C to 59% at 1000°C, or manufacture of hydrogen.

Low-pressure liquid coolants allow all their heat to be delivered at high temperatures, since the temperature drop in heat exchangers is less than with gas coolants. Also, with a good margin between operating and boiling temperatures, passive cooling for decay heat is readily achieved.

The removal of passive decay heat is a vital feature of primary cooling systems, beyond heat transfer to do work. When the fission process stops, fission product decay continues and a substantial amount of heat is added to the core. At the moment of shutdown, this is about 6% of the full power level, but it quickly drops to about 1% as the short-lived fission products decay. This heat could melt the core of a light water reactor unless it is reliably dissipated. Typically some kind of convection flow is relied upon.

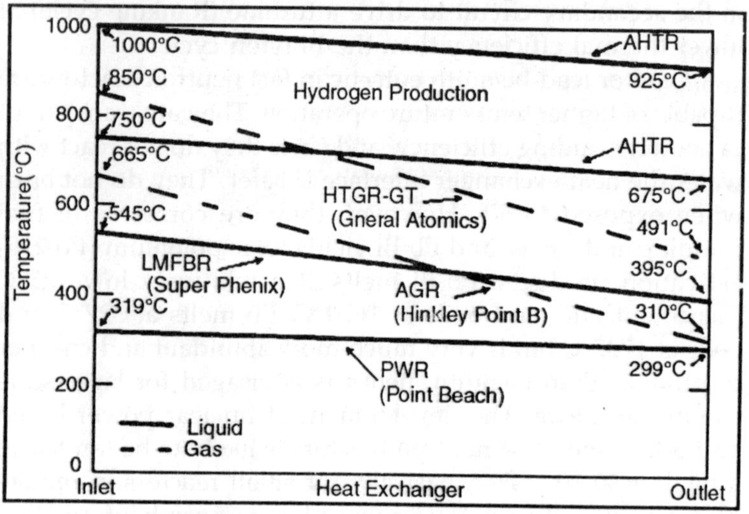

Fig. Heat Transfer for Different Primary Coolants Low-pressure Liquid Coolants Allow more Heat to be Delivered, at Higher Temperatures

Primitive Reactors

The world's oldest known nuclear reactors operated at what is now Oklo in Gabon, West Africa. About 2 billion years ago, at least 17 natural nuclear reactors achieved criticality in a rich deposit of uranium ore. Each operated at about 20 kW thermal. At that time the concentration of U-235 in all natural uranium was 3.7 per cent instead of 0.7 per cent as at present. (U-235 decays much faster than U-238, whose half-life is about the same as the age of the Earth.) These natural chain reactions, started spontaneously by the presence of water acting as a moderator, continued for about 2 million years before finally dying away.

During this long reaction period about 5.4 tonnes of fission products as well as 1.5 tonnes of plutonium together with other transuranic elements were generated in the orebody. The initial radioactive products have long since decayed into stable elements but close study of the amount and location of these has shown that there was little movement of radioactive wastes during and after the nuclear reactions. Plutonium and the other transuranics remained immobile.

Chapter 9

Conductors

CLASSIFICATION OF CONDUCTORS

By now you should be well aware of the correlation between electrical conductivity and certain types of materials. Those materials allowing for easy passage of free electrons are called conductors, while those materials impeding the passage of free electrons are called insulators.

Unfortunately, the scientific theories explaining why certain materials conduct and others don't are quite complex, rooted in quantum mechanical explanations in how electrons are arranged around the nuclei of atoms.

Contrary to the well-known"planetary" model of electrons whirling around an atom's nucleus as well-defined chunks of matter in circular or elliptical orbits, electrons in"orbit" don't really act like pieces of matter at all. Rather, they exhibit the characteristics of both particle and wave, their behaviour constrained by placement within distinct zones around the nucleus referred to as"shells" and"subshells."

Electrons can occupy these zones only in a limited range of energies depending on the particular zone and how occupied that zone is with other electrons.

If electrons really did act like tiny planets held in orbit around the nucleus by electrostatic attraction, their actions described by the same laws describing the motions of real planets, there could be no real distinction between conductors and insulators, and chemical bonds between atoms would not exist in the way they do now.

It is the discrete,"quantitized" nature of electron energy

and placement described by quantum physics that gives these phenomena their regularity.

When an electron is free to assume higher energy states around an atom's nucleus (due to its placement in a particular"shell"), it may be free to break away from the atom and comprise part

of an electric current through the substance.

If the quantum limitations imposed on an electron deny it this freedom, however, the electron is considered to be"bound" and cannot break away (at least not easily) to constitute a current. The former scenario is typical of conducting materials, while the latter is typical of insulating materials.

Some textbooks will tell you that an element's conductivity or nonconductivity is exclusively determined by the number of electrons residing in the atoms' outer"shell" (called the valence shell), but this is an oversimplification, as any examination of conductivity versus valence electrons in a table of elements will confirm.

The true complexity of the situation is further revealed when the conductivity of molecules (collections of atoms bound to one another by electron activity) is considered.

A good example of this is the element carbon, which comprises materials of vastly differing conductivity: graphite and diamond.

Graphite is a fair conductor of electricity, while diamond is practically an insulator (stranger yet, it is technically classified as a semiconductor, which in its pure form acts as an insulator, but can conduct under high temperatures and/ or the influence of impurities).

Both graphite and diamond are composed of the exact same types of atoms: carbon, with 6 protons, 6 neutrons and 6 electrons each.

The fundamental difference between graphite and diamond being that graphite molecules are flat groupings of carbon atoms while diamond molecules are tetrahedral (pyramid-shaped) groupings of carbon atoms.

If atoms of carbon are joined to other types of atoms to

form compounds, electrical conductivity becomes altered once again. Silicon carbide, a compound of the elements silicon and carbon, exhibits nonlinear behaviour: its electrical resistance decreases with increases in applied voltage! Hydrocarbon compounds (such as the molecules found in oils) tend to be very good insulators. As you can see, a simple count of valence electrons in an atom is a poor indicator of a substance's electrical conductivity.

All metallic elements are good conductors of electricity, due to the way the atoms bond with each other. The electrons of the atoms comprising a mass of metal are so uninhibited in their allowable energy states that they float freely between the different nuclei in the substance, readily motivated by any electric field.

The electrons are so mobile, in fact, that they are sometimes described by scientists as an electron gas, or even an electron sea in which the atomic nuclei rest. This electron mobility accounts for some of the other common properties of metals: good heat conductivity, malleability and ductility (easily formed into different shapes), and a lustrous finish when pure.

Thankfully, the physics behind all this is mostly irrelevant to our purposes here. Suffice it to say that some materials are good conductors, some are poor conductors, and some are in between. For now it is good enough to simply understand that these distinctions are determined by the configuration of the electrons around the constituent atoms of the material.

An important step in getting electricity to do our bidding is to be able to construct paths for electrons to flow with controlled amounts of resistance.

It is also vitally important that we be able to prevent electrons from flowing where we don't want them to, by using insulating materials. However, not all conductors are the same, and neither are all insulators. We need to understand some of the characteristics of common conductors and insulators, and be able to apply these characteristics to specific applications.

Almost all conductors possess a certain, measurable

resistance (special types of materials calledsuperconductors possess absolutely no electrical resistance, but these are not ordinary materials, and they must be held in special conditions in order to be super conductive).

Typically, we assume the resistance of the conductors in a circuit to be zero, and we expect that current passes through them without producing any appreciable voltage drop. In reality, however, there will almost always be a voltage drop along the (normal) conductive pathways of an electric circuit, whether we want a voltage drop to be there or not:

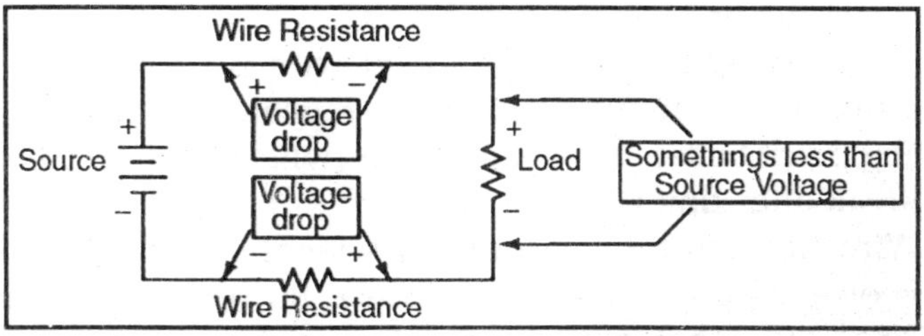

In order to calculate what these voltage drops will be in any particular circuit, we must be able to ascertain the resistance of ordinary wire, knowing the wire size and diameter.

REVIEW

- Electrical conductivity of a material is determined by the configuration of electrons in that materials atoms and molecules (groups of bonded atoms).
- All normal conductors possess resistance to some degree.
- Electrons flowing through a conductor with (any) resistance will produce some amount of voltage drop across the length of that conductor.

CONDUCTOR SIZE

It should be common-sense knowledge that liquids flow through large-diameter pipes easier than they do through small-diameter pipes (if you would like a practical illustration, try drinking a liquid through straws of different diameters). The same general principle holds for the flow of electrons through conductors: the broader the cross-sectional area (thickness) of the conductor, the more room for electrons to flow, and consequently, the easier it is for flow to occur (less resistance).

Electrical wire is usually round in cross-section (although there are some unique exceptions to this rule), and comes in two basic varieties: solid and stranded. Solid copper wire is just as it sounds: a single, solid strand of copper the whole length of the wire. Stranded wire is composed of smaller strands of solid copper wire twisted together to form a single, larger conductor.

The greatest benefit of stranded wire is its mechanical flexibility, being able to withstand repeated bending and twisting much better than solid copper (which tends to fatigue and break after time).

Wire size can be measured in several ways. We could speak of a wire's diameter, but since its really the cross-sectional area that matters most regarding the flow of electrons, we are better off designating wire size in terms of area.

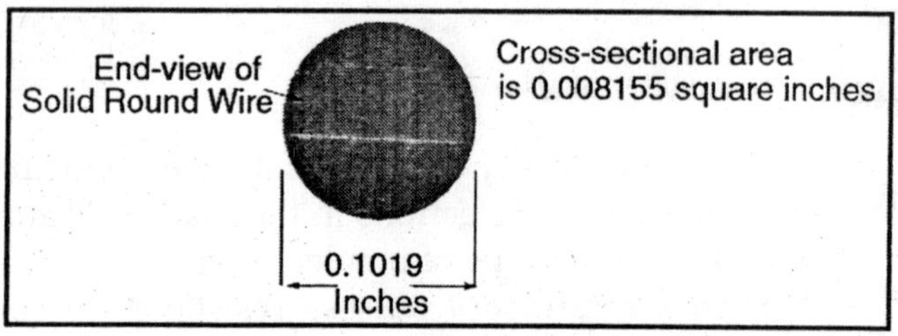

The wire cross-section picture shown above is, of course, not drawn to scale. The diameter is shown as being 0.1019 inches. Calculating the area of the cross-section with the formula Area = nr^2, we get an area of 0.008155 square inches:

$$A = \pi r^2$$

$$A = (3.1416)\left(\frac{0.1019\,\text{inches}}{2}\right)^2$$

$$A = 0.008155\,\text{square inches}$$

These are fairly small numbers to work with, so wire sizes

are often expressed in measures of thousandths-of-an-inch, or mils. For the illustrated example, we would say that the diameter of the wire was 101.9 mils (0.1019 inch times 1000). We could also, if we wanted, express the area of the wire in the unit of square mils, calculating that value with the same circle-area formula, Area = nr^2:

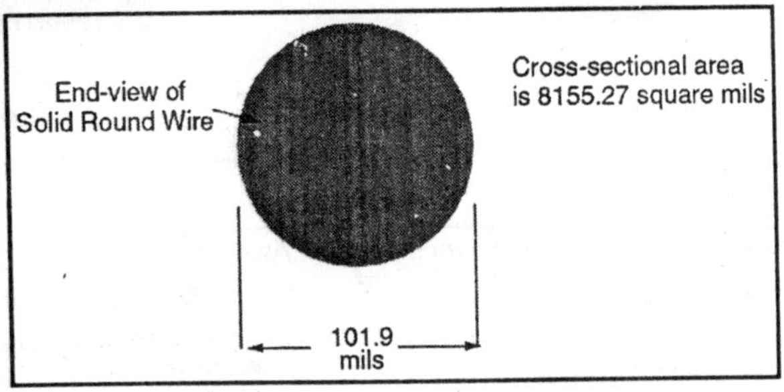

$$A = \pi r^2$$

$$A = (3.1416)\left(\frac{101.9\,\text{mils}}{2}\right)^2$$

$$A = 8155.27\,\text{square mils}$$

However, electricians and others frequently concerned with wire size use another unit of area measurement tailored specifically for wire's circular cross-section. This special unit is called the circular mil (sometimes abbreviated cmil). The sole purpose for having this special unit of measurement is to eliminate the need to invoke the factor n (3.1415927...) in the formula for calculating area, plus the need to figure wire radiuswhen you've been given diameter.

The formula for calculating the circular-mil area of a circular wire is very simple:

$$A=d^2$$

Because this is a unit of area measurement, the mathematical power of 2 is still in effect (doubling the width of a circle will always quadruple its area, no matter what units are used, or if the width of that circle is expressed in terms of radius or diameter). To illustrate the difference between measurements in square mils and measurements in circular mils, I will compare a circle with a square, showing the area of each shape in both unit measures:

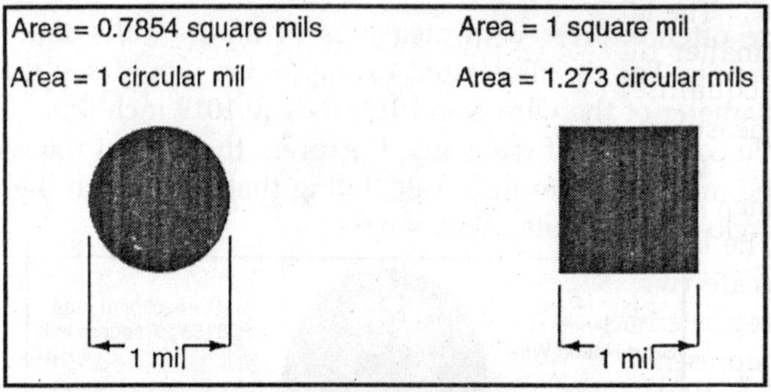

And for another size of wire:

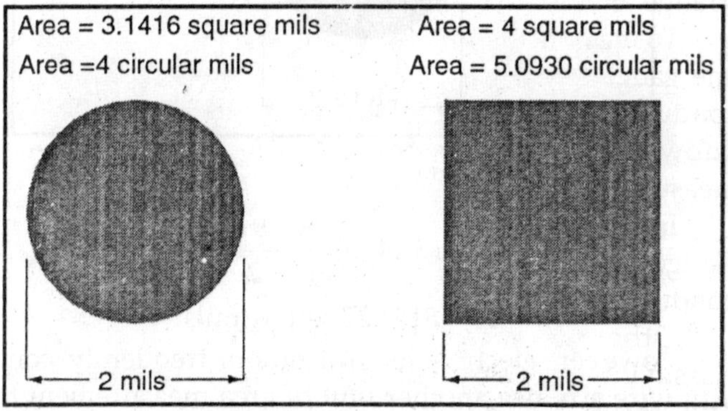

Obviously, the circle of a given diameter has less cross-sectional area than a square of width and height equal to the circle's diameter: both units of area measurement reflect that. However, it should be clear that the unit of"square mil" is really tailored for the convenient determination of a square's area, while"circular mil" is tailored for the convenient determination of a circle's area: the respective formula for each is simpler to work with.

It must be understood that both units are valid for measuring the area of a shape, no matter what shape that may be. The conversion between circular mils and square mils is a simple ratio: there are n (3.1415927) square mils to every 4 circular mils.

Another measure of cross-sectional wire area is the gauge. The gauge scale is based on whole numbers rather than fractional or decimal inches.

The larger the gauge number, the skinnier the wire; the smaller the gauge number, the fatter the wire. For those acquainted with shotguns, this inversely-proportional measurement scale should sound familiar.

The table at the end of this section equates gauge with inch diameter, circular mils, and square inches for solid wire. The larger sizes of wire reach an end of the common gauge scale (which naturally tops out at a value of 1), and are represented by a series of zeros."3/0" is another way to represent"000," and is pronounced"triple-ought." Again, those acquainted with shotguns should recognize the terminology, strange as it may sound.

To make matters even more confusing, there is more than one gauge"standard" in use around the world. For electrical conductor sizing, the American Wire Gauge (AWG), also known as the Brown and Sharpe (B&S) gauge, is the measurement system of choice.

In Canada and Great Britain, the British Standard Wire Gauge (SWG) is the legal measurement system for electrical conductors.

Other wire gauge systems exist in the world for classifying wire diameter, such as the Stubs steel wire gauge and the Steel Music Wire Gauge (MWG), but these measurement systems apply to non-electrical wire use.

The American Wire Gauge (AWG) measurement system, despite its oddities, was designed with a purpose: for every three steps in the gauge scale, wire area (and weight per unit length) approximately doubles. This is a handy rule to remember when making rough wire size estimations!

For very large wire sizes (fatter than 4/0), the wire gauge system is typically abandoned for cross-sectional area measurement in thousands of circular mils (MCM), borrowing the old Roman numeral"M" to denote a multiple of"thousand" in front of"CM" for"circular mils."

The following table of wire sizes does not show any sizes bigger than 4/0 gauge, because solid copper wire becomes impractical to handle at those sizes. Stranded wire construction is favoured, instead.

WIRE TABLE FOR SOLID, ROUND COPPER CONDUCTORS

Size AWG	Diameter inches	Cross-sectional area cir. mils	Cross-sectional area sq. inches	Weight lb/1000 ft
4/0	0.4600	211,600	0.1662	640.5
3/0	0.4096	167,800	0.1318	507.9
2/0	0.3648	133,100	0.1045	402.8
1/0	0.3249	105,500	0.08289	319.5
1	0.2893	83,690	0.06573	253.5
2	0.2576	66,370	0.05213	200.9
3	0.2294	52,630	0.04134	159.3
4	0.2043	41,740	0.03278	126.4
5	0.1819	33,100	0.02600	100.2
6	0.1620	26,250	0.02062	79.46
7	0.1443	20,820	0.01635	63.02
8	0.1285	16,510	0.01297	49.97
9	0.1144	13,090	0.01028	39.63
10	0.1019	10,380	0.008155	31.43
11	0.09074	8,234	0.006467	24.92
12	0.08081	6,530	0.005129	19.77
13	0.07196	5,178	0.004067	15.68
14	0.06408	4,107	0.003225	12.43
15	0.05707	3,257	0.002558	9.858
16	0.05082	2,583	0.002028	7.818
17	0.04526	2,048	0.001609	6.200
18	0.04030	1,624	0.001276	4.917
19	0.03589	1,288	0.001012	3.899
20	0.03196	1,022	0.0008023	3.092
21	0.02846	810.1	0.0006363	2.452
22	0.02535	642.5	0.0005046	1.945
23	0.02257	509.5	0.0004001	1.542
24	0.02010	404.0	0.0003173	1.233
25	0.01790	320.4	0.0002517	0.9699
26	0.01594	254.1	0.0001936	0.7692
27	0.01420	201.5	0.0001583	0.6100
28	0.01264	159.8	0.0001255	0.4837
29	0.01126	126.7	0.00009954	0.3836
30	0.01003	100.5	0.00007894	0.3042

31	0.008928	79.70	0.00006260	0.2413
32	0.007950	63.21	0.00004964	0.1913
33	0.007080	50.13	0.00003937	0.1517
34	0.006305	39.75	0.00003122	0.1203
35	0.005615	31.52	0.00002476	0.09542
36	0.005000	25.00	0.00001963	0.07567
37	0.004453	19.83	0.00001557	0.06001
38	0.003965	15.72	0.00001235	0.04759
39	0.003531	12.47	0.000009793	0.03774
40	0.003145	9.888	0.000007766	0.02993
41	0.002800	7.842	0.000006159	0.02374
42	0.002494	6.219	0.000004884	0.01882
43	0.002221	4.932	0.000003873	0.01493
44	0.001978	3.911	0.000003072	0.01184

For some high-current applications, conductor sizes beyond the practical size limit of round wire are required. In these instances, thick bars of solid metal called busbars are used as conductors.

Busbars are usually made of copper or aluminum, and are most often uninsulated. They are physically supported away from whatever framework or structure is holding them by insulator standoff mounts.

Although a square or rectangular cross-section is very common for busbar shape, other shapes are used as well. Cross-sectional area for busbars is typically rated in terms of circular mils (even for square and rectangular bars!), most likely for the convenience of being able to directly equate busbar size with round wire.

REVIEW

- Electrons flow through large-diameter wires easier than small-diameter wires, due to the greater cross-sectional area they have in which to move.
- Rather than measure small wire sizes in inches, the unit of"mil" (1/1000 of an inch) is often employed.
- The cross-sectional area of a wire can be expressed in terms of square units (square inches or square mils), circular mils, or"gauge" scale.

- Calculating square-unit wire area for a circular wire involves the circle area formula:
- $A=\pi r^2$(Square units)
- Calculating circular-mil wire area for a circular wire is much simpler, due to the fact that the unit of"circular mil" was sized just for this purpose: to eliminate the"pi" and the d/2 (radius) factors in the formula.
- $A=d^2$(Circular units)
- There are ? (3.1416) square mils for every 4 circular mils.
- The gauge system of wire sizing is based on whole numbers, larger numbers representing smaller-area wires and vice versa. Wires thicker than 1 gauge are represented by zeros: 0, 00, 000, and 0000 (spoken"single-ought,""double-ought,""triple-ought," and"quadruple-ought."
- Very large wire sizes are rated in thousands of circular mils (MCM's), typical for busbars and wire sizes beyond 4/0.
- Busbars are solid bars of copper or aluminum used in high-current circuit construction. Connections made to busbars are usually welded or bolted, and the busbars are often bare (uninsulated), supported away from metal frames through the use of insulating standoffs.

CONDUCTOR AMPACITY

The smaller the wire, the greater the resistance for any given length, all other factors being equal. A wire with greater resistance will dissipate a greater amount of heat energy for any given amount of current, the power being equal to P=I2R.

Dissipated power in a resistance manifests itself in the form of heat, and excessive heat can be damaging to a wire (not to mention objects near the wire!), especially considering the fact that most wires are insulated with a plastic or rubber coating, which can melt and burn. Thin wires will, therefore, tolerate less current than thick wires, all other factors being equal. A conductor's current-carrying limit is known as its ampacity.

Primarily for reasons of safety, certain standards for electrical wiring have been established within the United States, and are specified in the National Electrical Code (NEC). Typical NEC wire ampacity tables will show allowable maximum currents for different sizes and applications of wire. Though the melting point of copper theoretically imposes a limit on wire ampacity, the materials commonly employed for insulating conductors melt at temperatures far below the melting point of copper, and so practical ampacity ratings are based on the thermal limits of the insulation. Voltage dropped as a result of excessive wire resistance is also a factor in sizing conductors for their use in circuits, but this consideration is better assessed through more complex means. A table derived from an NEC listing is shown for example:

COPPER CONDUCTOR AMPACITIES, IN FREE AIR AT 30 DEGREES C

INSULATION TYPE:	RUW, T TW	THW, THWN RUH	FEP, FEPB THHN, XHHW
Size AWG	Current Rating @ 60 degrees C	Current Rating @ 75 degrees C	Current Rating @ 90 degrees C
20	9		12.5
18	13		18
16	18		24
14	25	30	35
12	30	35	40
10	40	50	55
8	60	70	80
6	80	95	105
4	105	125	140
2	140	170	190
1	165	195	220
1/0	195	230	260
2/0	225	265	300
3/0	260	310	350
4/0	300	360	405

Notice the substantial ampacity differences between same-size wires with different types of insulation. This is due, again, to the thermal limits (60o, 75o, 90o) of each type of insulation material.

These ampacity ratings are given for copper conductors in"free air" (maximum typical air circulation), as opposed to wires placed in conduit or wire trays. As you will notice, the table fails to specify ampacities for small wire sizes. This is because the NEC concerns itself primarily with power wiring (large currents, big wires) rather than with wires common to low-current electronic work.

There is meaning in the letter sequences used to identify conductor types, and these letters usually refer to properties of the conductor's insulating layer(s). Some of these letters symbolize individual properties of the wire while others are simply abbreviations. For example, the letter"T" by itself means"thermoplastic" as an insulation material, as in"TW" or"THHN." However, the three-letter combination"MTW" is an abbreviation forMachine Tool Wire, a type of wire whose insulation is made to be flexible for use in machines experiencing significant motion or vibration.

```
INSULATION MATERIAL:
C = Cotton
FEP = Fluorinated Ethylene Propylene
MI = Mineral (magnesium oxide)
PFA = Perfluoroalkoxy
R = Rubber (sometimes Neoprene)
S = Silicone"rubber"
SA = Silicone-asbestos
T = Thermoplastic
TA = Thermoplastic-asbestos
TFE = Polytetrafluoroethylene ("Teflon")
X = Cross-linked synthetic polymer
Z = Modified ethylene tetrafluoroethylene
HEAT RATING:
H = 75 degrees Celsius
HH = 90 degrees Celsius
OUTER COVERING ("JACKET"):
```

```
N = Nylon
SPECIAL SERVICE CONDITIONS:
U = Underground
W = Wet
-2 = 90 degrees Celsius and wet
```

Therefore, a"THWN" conductor has Thermoplastic insulation, is Heat resistant to 75o Celsius, is rated forWet conditions, and comes with a Nylon outer jacketing.

Letter codes like these are only used for general-purpose wires such as those used in households and businesses. For high-power applications and/or severe service conditions, the complexity of conductor technology defies classification according to a few letter codes. Overhead power line conductors are typically bare metal, suspended from towers by glass, porcelain, or ceramic mounts known as insulators.

Even so, the actual construction of the wire to withstand physical forces both static (dead weight) and dynamic (wind) loading can be complex, with multiple layers and different types of metals wound together to form a single conductor. Large, underground power conductors are sometimes insulated by paper, then enclosed in a steel pipe filled with pressurized nitrogen or oil to prevent water intrusion. Such conductors require support equipment to maintain fluid pressure throughout the pipe. Other insulating materials find use in small-scale applications. For instance, the small-diameter wire used to make electromagnets (coils producing a magnetic field from the flow of electrons) are often insulated with a thin layer of enamel. The enamel is an excellent insulating material and is very thin, allowing many"turns" of wire to be wound in a small space.

REVIEW

- Wire resistance creates heat in operating circuits. This heat is a potential fire ignition hazard.
- Skinny wires have a lower allowable current ("ampacity") than fat wires, due to their greater resistance per unit length, and consequently greater heat generation per unit current.

- The National Electrical Code (NEC) specifies ampacities for power wiring based on allowable insulation temperature and wire application.

INSULATORS AND SEMICONDUCTORS

In this study of electricity and electronics, the association of matter and electricity is important. Since every electronic device is constructed of parts made from ordinary matter, the effects of electricity on matter must be well understood. As a means of accomplishing this, all elements of which matter is made may be placed into one of three categories: CONDUCTORS, SEMICONDUCTORS, and INSULATORS, depending on their ability to conduct an electric current. CONDUCTORS are elements which conduct electricity very readily, INSULATORS have an extremely high resistance to the flow of electricity. All matter between these two extremes may be called SEMICONDUCTORS.

The electron theory states that all matter is composed of atoms and the atoms are composed of smaller particles called protons, electrons, and neutrons. The electrons orbit the nucleus which contains the protons and neutrons. It is the valence electrons that we are most concerned with in electricity. These are the electrons which are easiest to break loose from their parent atom. Normally, conductors have three or less valence electrons; insulators have five or more valence electrons; and semiconductors usually have four valence electrons.

The electrical conductivity of matter is dependent upon the atomic structure of the material from which the conductor is made. In any solid material, such as copper, the atoms which make up the molecular structure are bound firmly together. At room temperature, copper will contain a considerable amount of heat energy. Since heat energy is one method of removing electrons from their orbits, copper will contain many free electrons that can move from atom to atom.

When not under the influence of an external force, these electrons move in a haphazard manner within the conductor. This movement is equal in all directions so that electrons are not lost or gained by any part of the conductor. When controlled by an external force, the electrons move generally in the same direction. The effect of this movement is felt almost instantly from one end of the conductor to the other. This electron movement is called an ELECTRIC CURRENT.

Some metals are better conductors of electricity than others. Silver, copper, gold, and aluminum are materials with many free electrons and make good conductors. Silver is the best conductor, followed by copper, gold, and aluminum. Copper is used more often than silver because of cost. Aluminum is used where weight is a major consideration, such as in high-tension power lines, with long spans between supports. Gold is used where oxidation or corrosion is a consideration and a good conductivity is required. The ability of a conductor to handle current also depends upon its physical dimensions. Conductors are usually found in the form of wire, but may be in the form of bars, tubes, or sheets.

Nonconductors have few free electrons. These materials are called INSULATORS. Some examples of these materials are rubber, plastic, enamel, glass, dry wood, and mica. Just as there is no perfect conductor, neither is there a perfect insulator.

Some materials are neither good conductors nor good insulators, since their electrical characteristics fall between those of conductors and insulators. These in-between materials are classified as SEMICONDUCTORS. Germanium and silicon are two common semiconductors used in solid-state devices.

ELECTROSTATICS

Electrostatics (electricity at rest) is a subject with which most persons entering the field of electricity and electronics are somewhat familiar. For example, the way a person's hair stands on end after a vigourous rubbing is an effect of electrostatics. While pursuing the study of electrostatics, you will gain a better understanding of this common occurrence. Of even greater significance, the study of electrostatics will provide you with the opportunity to gain important background knowledge and to develop concepts which are essential to the understanding of electricity and electronics.

Interest in the subject of static electricity can be traced back to the Greeks. Thales of Miletus, a Greek philosopher and mathematician, discovered that when an amber rod is rubbed with fur, the rod has the amazing characteristic of attracting some very light objects such as bits of paper and shavings of wood.

About 1600, William Gilbert, an English scientist, made a study of other substances which had been found to possess qualities of attraction similar to amber. Among these were glass, when rubbed with silk, and ebonite, when rubbed with fur. Gilbert classified all the substances which possessed properties similar to those of amber as electrics, a word of Greek origin meaning amber.

Because of Gilbert's work with electrics, a substance such as amber or glass when given a vigourous rubbing was recognized as being ELECTRIFIED, or CHARGED with electricity.

In the year 1733, Charles Dufay, a French scientist, made an important discovery about electrification. He found that when a glass was rubbed with fur, both the glass rod and the fur became electrified. This realization came when he systematically placed the glass rod and the fur near other electrified substances and found that certain substances which were attracted to the glass rod were repelled by the fur, and vice versa. From experiments such as this, he concluded that there must be two exactly opposite kinds of electricity.

Benjamin Franklin, American statesman, inventor, and philosopher, is credited with first using the terms POSITIVE and NEGATIVE to describe the two opposite kinds of electricity. The charge produced on a glass rod when it is rubbed with silk, Franklin labeled positive. He attached the term negative to the charge produced on the silk. Those bodies which were not electrified or charged, he called NEUTRAL.

STATIC ELECTRICITY

In a natural, or neutral state, each atom in a body of

matter will have the proper number of electrons in orbit around it. Consequently, the whole body of matter composed of the neutral atoms will also be electrically neutral. In this state, it is said to have a"zero charge." Electrons will neither leave nor enter the neutrally charged body should it come in contact with other neutral bodies. If, however, any number of electrons are removed from the atoms of a body of matter, there will remain more protons than electrons and the whole body of matter will become ELECTRICALLY POSITIVE. Should the positively charged body come in contact with another body having a normal charge, or having a NEGATIVE (too many electrons) charge, an electric current will flow between them. Electrons will leave the more negative body and enter the positive body. This electron flow will continue until both bodies have equal charges. When two bodies of matter have unequal charges and are near one another, an electric force is exerted between them because of their unequal charges. However, since they are not in contact, their charges cannot equalize. The existence of such an electric force, where current cannot flow, is referred to as static electricity. ("Static" in this instance means"not moving.") It is also referred to as an electrostatic force.

One of the easiest ways to create a static charge is by friction. When two pieces of matter are rubbed together, electrons can be"wiped off" one material onto the other. If the materials used are good conductors, it is quite difficult to obtain a detectable charge on either, since equalizing currents can flow easily between the conducting materials. These currents equalize the charges almost as fast as they are created. A static charge is more easily created between nonconducting materials. When a hard rubber rod is rubbed with fur, the rod will accumulate electrons given up by the fur, as shown in. Since both materials are poor conductors, very little equalizing current can flow, and an electrostatic charge builds up. When the charge becomes great enough, current will flow regardless of the poor conductivity of the materials. These currents will cause visible sparks and produce a crackling sound.

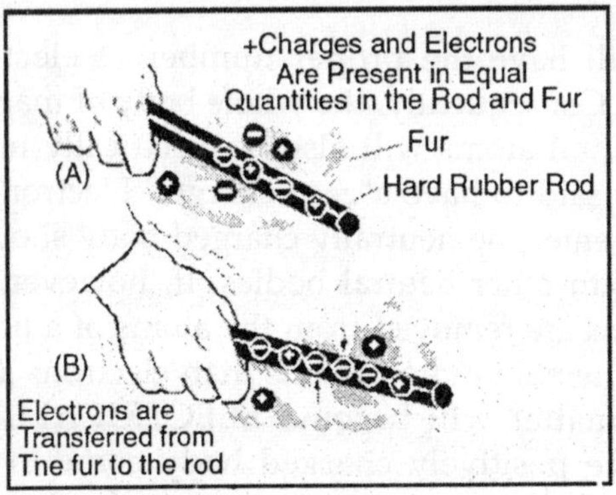

NATURE OF CHARGES

When in a natural, or neutral state, an atom has an equal number of electrons and protons. Because of this balance, the net negative charge of the electrons in orbit is exactly balanced by the net positive charge of the protons in the nucleus, making the atom electrically neutral. An atom becomes a positive ion whenever it loses an electron, and has an overall positive charge. Conversely, whenever an atom acquires an extra electron, it becomes a negative ion and has a negative charge.

Due to normal molecular activity, there are always ions present in any material. If the number of positive ions and negative ions is equal, the material is electrically neutral. When the number of positive ions exceeds the number of negative ions, the material is positively charged. The material is negatively charged whenever the negative ions outnumber the positive ions. Since ions are actually atoms without their normal number of electrons, it is the excess or the lack of electrons in a substance that determines its charge. In most solids, the transfer of charges is by movement of electrons rather than ions. The transfer of charges by ions will become more significant when we consider electrical activity in liquids and gases. At this time, we will discuss electrical behaviour in terms of electron movement.

CHARGED BODIES

One of the fundamental laws of electricity is that LIKE

CHARGES REPEL EACH OTHER and UNLIKE CHARGES ATTRACT EACH OTHER. A positive charge and negative charge, being unlike, tend to move toward each other. In the atom, the negative electrons are drawn toward the positive protons in the nucleus.

This attractive force is balanced by the electron's centrifugal force caused by its rotation about the nucleus. As a result, the electrons remain in orbit and are not drawn into the nucleus. Electrons repel each other because of their like negative charges, and protons repel each other because of their like positive charges. The law of charged bodies may be demonstrated by a simple experiment. Two paper pulp balls are suspended near one another by threads.

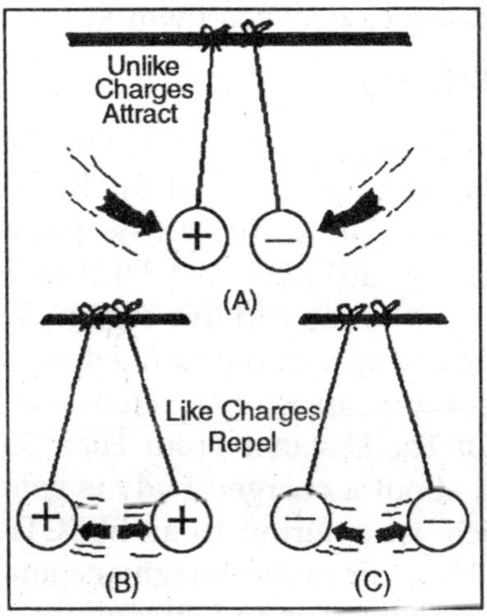

If a hard rubber rod is rubbed with fur to give it a negative charge and is then held against the right-hand ball in (A), the rod will give off a negative charge to the ball. The right-hand ball will have a negative charge with respect to the left-hand ball.

When released, the two balls will be drawn together, (A). They will touch and remain in contact until the left-hand ball gains a portion of the negative charge of the right-hand ball, at which time they will swing apart (C). If a positive or a negative charge is placed on both balls (B), the balls will repel each other.

COULOMB'S LAW OF CHARGES

The relationship between attracting or repelling charged bodies was first discovered and written about by a French scientist named Charles A. Coulomb. Coulomb's Law states that Charged Bodies Attract Or Repel Each Other With A Force That Is Directly Proportional To The Product Of Their Individual Charges, And Is Inversely Proportional To The Square Of The Distance Between Them. The amount of attracting or repelling force which acts between two electrically charged bodies in free space depends on two things:

- Their charges and
- The distance between them.

ELECTRIC FIELDS

The space between and around charged bodies in which their influence is felt is called an ELECTRIC FIELD OF FORCE. It can exist in air, glass, paper, or a vacuum. ELECTROSTATIC FIELDS and DIELECTRIC FIELDS are other names used to refer to this region of force.

Fields of force spread out in the space surrounding their point of origin and, in general, Diminish In Proportion To The Square Of The Distance From Their Source.

The field about a charged body is generally represented by lines which are referred to as ELECTROSTATIC LINES OF FORCE. These lines are imaginary and are used merely to represent the direction and strength of the field. To avoid confusion, the lines of force exerted by a positive charge are always shown leaving the charge, and for a negative charge they are shown entering. illustrates the use of lines to represent the field about charged bodies.

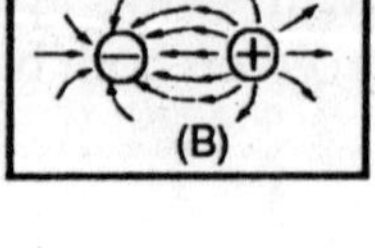

(A) Represents the repulsion of like-charged bodies and their associated fields.

(B) Represents the attraction of unlike-charged bodies and their associated fields.

ENERGY BANDS IN SOLIDS

Atoms are clustered together and are overlapped in solids. Therefore, the outermost valence atoms are overlapped. No two atoms can have the same quantum state. Hence, all electrons have discrete energy states. If there are'n' orbits, then there are 2n possible energy levels which are splitting and creating a band of energy.

VALENCE BAND

When the electrons have closely spaced energy levels, a valence band is created.

Energy Gap

Valence band and the conduction band lies in the energy gap. According to the Bohr's theory, free electrons in an isolated atom have certain definite discrete amount of energy. If large number of atoms are brought close to one another to form a crystal, they begin to influence each other. The valence electrons are attracted by the nucleus of the other atoms. This brings about a considerable modification in the case of energy levels of the electrons in the outer shells. *The process of splitting of energy levels can be understood as follows*:

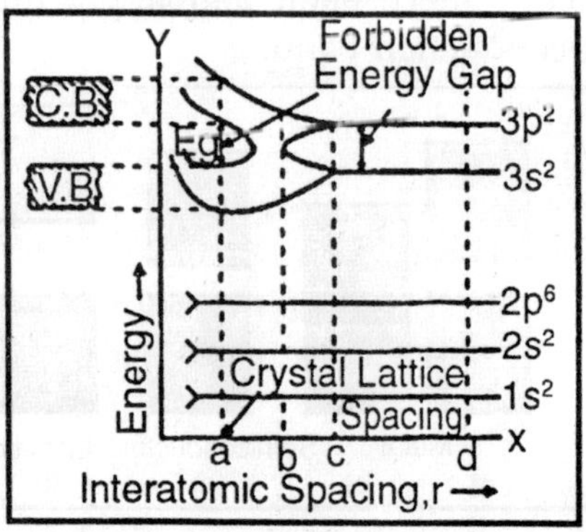

- If interatomic spacing of atoms is very large i.e., r =

d>>a, there is no interatomic separation. Each atom in the crystal behaves as free atom. Take for example silicon whose electronic configuration is $1s^2\ 2s^2\ 2p^6\ 3s^2\ 3p^2$. If N atoms were to be considered in silicon crystal, then there will be 2N electrons filling 2N possible energy levels in 3s, 6N possible levels in 3p of which only 2N is completely filled.

- When the spacing is progressively decreased i.e., c<r<d, there is no visible splitting of energy levels.
- When r = c, the 3s and 3p electrons of neighbouring silicon atoms becomes appreciable. The energy of electrons of each atom starts changing, whereas the energies of electrons in the inner shell do not change.
- When r lies between b and c, the energy levels get slightly changed and instead of a single 3s or 3p levels, we get a large number of closely packed levels. This collection of closely spaced energy levels is called an energy band.
- When r = b>a, the gap between 3s and 3p completely disappear and the 8N energy levels are (2N of 3s and 6N of 3p sub shells) continuously distributed. In this stage 4N levels are filled and 4N levels are empty.
- When r = a i.e., actual spacing in the crystal the 4N filled energy levels are separated from 4N unfilled energy levels. This gap or separation is called the forbidden gap. E.g., the lower completely filled band is called valence band and upper unfilled band is called conduction band.

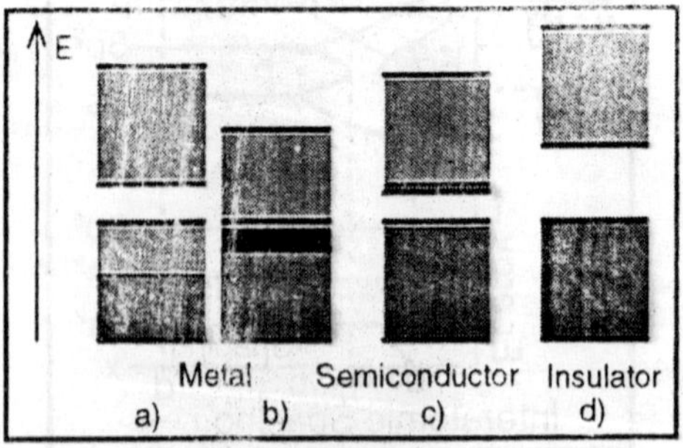

Fig. Energy bands in solids

PN JUNCTION

We've seen that it is possible to turn a crystal of pure silicon into a moderately good electrical conductor by adding an impurity such as arsenic or phosphorus (for an N-type semiconductor) or aluminum or gallium (for a P-type semiconductor). By itself, however, a single type of semiconductor material isn't very useful. Useful applications start to happen only when a single semiconductor crystal contains both P-type and N-type regions. Here we will examine the properties of a single silicon crystal which is half N-type and half P-type.

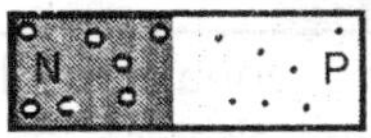

Consider the silicon crystal represented to the right. Half is N-type while the other half is P-type. We've shown the two types separated slightly, as if they were two separate crystals. The free electrons in the N-type crystal are represented by small black circles with a"-" sign inside to indicate their polarity. The holes in the P-type crystal are shown as small white circles with a"+" inside.

In the real world, it isn't possible to join two such crystals together usefully. Therefore, a practical PN junction can only be created by inserting different impurities into different parts of a single crystal. So let's see what happens when we join the N- and P-type crystals together, so that the result is one crystal with a sharp boundary between the two types.

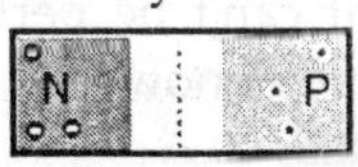

You might think that, left to itself, it would just sit there. However, this is not the case. Instead, an interesting interaction occurs at the junction. The extra electrons in the N region will seek to lose energy by filling the holes in the P region. This leaves an empty zone, or depletion region as it is called, around the junction as shown to the right. This action also leaves a small electrical imbalance inside the crystal. The N region is missing some electrons so it has a positive charge. Those electrons have migrated to fill holes in the P region, which therefore has a negative charge. This electrical imbalance amounts to about 0.3 volt in a germanium crystal, and about 0.65 to 0.7 volt in a silicon crystal. This will vary somewhat depending on the concentration of the impurities on either side of the junction.

Unfortunately, it is not possible to exploit this electrical imbalance as a power source; it doesn't work that way. However, we can apply an external voltage to the crystal and see what happens in response. Let's take a look at the possibilities.

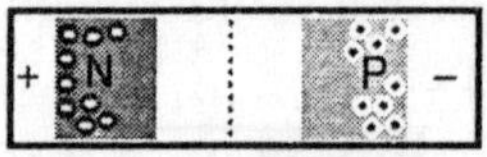

Suppose we apply a voltage to the outside ends of our PN crystal. We have two choices. In this case, the positive voltage is applied to the N-type material. In response, we see that the positive voltage applied to the N-type material attracts any free electrons towards the end of the crystal and away from the junction, while the negative voltage applied to the P-type end attracts holes away from the junction on this end.

The result is that all available current carriers are attracted away from the junction, and the depletion region grows correspondingly larger.

There is no current flow through the crystal because all available current carriers are attracted away from the junction, and cannot cross. (We are here considering an ideal crystal -- in real life, the crystal can't be perfect, and some leakage current does flow.) This is known as reverse bias applied to the semiconductor crystal.

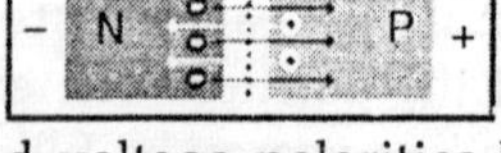

Here the applied voltage polarities have been reversed. Now, the negative volatge applied to the N-type end pushes electrons towards the junction, while the positive voltage at the P-type end pushes holes towards the junction. This has the effect of shrinking the depletion region.

As the applied voltage exceeds the internal electrical

imbalance, current carriers of both types can cross the junction into the opposite ends of the crystal. Now, electrons in the P-type end are attracted to the positive applied voltage, while holes in the N-type end are attracted to the negative applied voltage. This is the condition of forward bias.

Because of this behaviour, an electrical current can flow through the junction in the forward direction, but not in the reverse direction. This is the basic nature of an ordinary semicond ictor diode.

It is important to realise that holes exist only within the crystal. A hole reaching the negative terminal of the crystal is filled by an electron from the power source and simply disappears. At the positive terminal, the power supply attracts an electron out of the crystal, leaving a hole behind to move through the crystal toward the junction again.

In some literature, you might see the N-type connection designated the cathode of the diode, while the P-type connection is called the anode. These designations come from the days of vacuum tubes, but are still in use. Electrons always move from cathode to anode inside the diode.

One point that needs to be recognized is that there is a limit to the magnitude of the reverse voltage that can be applied to any PN junction. As the applied reverse voltage increases, the depletion region continues to expand. If either end of the depletion region approaches its electrical contact too closely, the applied voltage has become high enough to generate an electrical arc straight through the crystal. This will destroy the diode.

It is also possible to allow too much current to flow through the diode in the forward direction. The crystal is not a perfect conductor, remember; it does exhibit some resistance. Heavy current flow will generate some heat within that resistance. If the resulting temperature gets too high, the semiconductor crystal will actually melt, again destroying its usefulness.

Always be careful to pay attention to the maximum specifications of a diode, and be sure to keep the operating conditions of the diode well within the indicated limits.

PN DIODE

The most basic property of a junction diode is that it conducts an electric current in one direction and blocks it in the other. This behaviour arises from the electrical characteristics of a junction, called a p-n junction.fabricated within a semiconductor crystal. The most commonly used semiconductor material is silicon. The junction diode is useful in a wide variety of applications including the rectification of ac signals (producing dc from ac), the detection of radio signals, the conversion of solar power to electricity, and in the generation and detection of light.

It also finds use in a variety of electronic circuits as a switch, as a voltage reference or even as a tunable capacitor. The p-n junction is also the basic building block of a host of other electronic devices, of which the most well-known is the junction transistor. For this reason, a study of the properties and behaviour of the p-n junction is important.

In this chapter, the conduction of electricity in solids is reviewed first and the conduction properties of semiconductors are then explained. The construction of the p-n junction of an ideal diode is described and an explanation of the operation of the device is presented.

CONDUCTION IN SOLIDS

All matter consists of atoms. Each atom has electrons orbiting the nucleus. The nucleus contains the same amount of positive charge as the negative charge possessed by the orbiting electrons. The ability of any material to conduct electricity depends primarily on the behaviour of the electrons in the outer orbits. Therefore, it is necessary to review briefly some aspects of solid-state physics. This subject will be dealt with in more detail later in the course.

CONDUCTORS

In a metallic conductor such as copper, the atoms are arranged in a regular array called a crystal lattice. The electrons in the outer orbits of each metal atom are only loosely bound to the nucleus. These electrons are not closely associated with any particular atom and are free to move through the crystal lattice. Once an electron has left its orbit around a particular atom, that atom is left with an excess positive charge. The electron-deficient atom is called a positive ion. The electron that is now free to move is called a free electron. The free electrons in a conductor can be visualized as a cloud of electrons surrounding fixed positive ions as shown in.

At normal temperatures, the ions possess energy and vibrate. Collisions between vibrating ions and free electrons cause the electrons to move in a random manner. Over a long period of time, the net motion of these free electrons is zero.

If an electric field is applied to the conductor, the free electrons will acquire additional energy and will tend to move in the direction dictated by the field. There will be a resulting net motion of free electrons. The net motion of charge carriers constitutes an electric current.

INSULATORS

In an insulator, nearly all electrons are very tightly bound to their respective atoms. There are practically no electrons that are able to move under the influence of an applied electric field. Therefore, an insulator cannot conduct any appreciable electric current under normal conditions.

SEMICONDUCTORS

A semiconductor, such as silicon, has properties somewhere between those of a conductor and an insulator. The ability of a semiconductor to conduct electricity can be changed dramatically by adding small numbers of a different element to the semiconductor crystal. This process is called doping. Early experiments showed that an electric current through a semiconductor was carried by the flow of positivecharges as well as negative charges (electrons).

DOPED SEMICONDUCTORS

A semiconductor crystal is called n-type if the addition of an impurity element results in a large number if free electrons (negative charge carriers) available for conduction. Each impurity atom is called a donoratom since it donates an electron. The electron is free to move and can contribute to an electric current. The positive ion left behind is fixed and cannot take part in conduction.

A semiconductor crystal can be made p-type by doping it with a different element so that there are a large number of positive charge carriers available for conduction. The positive charge carriers actually correspond to vacancies or deficiencies of electrons in the bonds holding the atoms in the crystal lattice. The positive charges are called holes. These holes can move through the lattice as illustrated in one dimension in (a) and (b). The dotted lines represent the crystal lattice.

Note that the movement of a hole is due to the movement of a bound electron from one bond to another. It is not due to the motion of free electrons.

In a p-type semiconductor, most of the mobile charge carriers are holes. A hole moving away from its host impurity atom is equivalent to the atom gaining or accepting an electron into its bonding structure. The host atom gains an excess negative charge and is then called an acceptor ion. This situation is illustrated in.

Note again that the ions are locked in the crystal lattice and therefore reperesent fixed charges and cannot contribute to current. On the other hand, the holes are mobile charge carriers and can contribute to current flow.

Even in a highly doped p-type semiconductor there will always be some free electrons. This very small number of free electrons have been omitted in Figure 4 for clarity. Similarly, n-type semiconductors always contain some holes. The predominant mobile charge carriers are called majority carriers, whilst those in the minority are called minority carriers. For example, the majority carriers in n-type material are free electrons.

The terms majority carriers and minority carriers have meaning only if the type of semiconductor (n- or p-) is specified. A pure or undoped semiconductor is said to be intrinsic. Such material has equal numbers of holes and free electrons. These carriers are produced as a result of thermal agitation of the atoms, even at room temperature. Some bound electrons can acquire sufficient energy to escape from their atoms, becoming free electrons and leaving holes behind., This process of producing hole-electron pairs is called thermal generation.

It is possible for a free electron and a hole to come near each other in the course of their random wandering through the crystal. The free electron can then occupy the vacant position represented by the hole. The hole and electron are said to recombine. There is then no mobile charge carrier at that point. The rate of recombination depends upon the number of carriers present.

Thermal generation and recombination occurs in both doped and undoped semiconductor material. When a semiconductor material is in thermal equilibrium, the rate of generation of hole-electron pairs equals the rate of recombination. The density or concentration of both holes and electrons then remains constant.

CURRENT FLOW IN SEMICONDUCTORS

An electric current can flow through a semiconductor as a result of the movement of holes and/or free electrons. There are two important processes that account for current flow in semiconductors. These processes are called drift and diffusion.

DRIFT

Applying an electric field across a semiconductor will cause holes and free electrons to drift through the crystal in the directions The total current is equal to the sum of hole current (to the right) and electron current (tpo the left).

DIFFUSION

A drop of ink in a glass of water diffuses through the water until it is evenly distributed. The same process, called diffusion, occurs with semiconductors. For example, if some extra free electrons are introduced into a p-type semiconductor, the free electrons will redistribute themselves so that the concentration is more uniform. In the example, the free electrons will tend to move to the right. This net motion of charge carriers constitutes a diffusion current.(the holes and acceptor ions are omitted for clarity.)

In this example, the free electrons move away from the region of highest concentration. The higher the localized concentration, the greater will be the rate at which electrons move away. The same process applies to holes in an n-type semiconductor.

Note that when a few minority carriers are diffusing through a sample, they will encounter a large number of majority carriers. Some recombination will occur. A number of both types of carrier will be lost.

THE P-N JUNCTION

Imagine that a p-type block of silicon can be placed in perfect contact with an n-type block. Free electrons from the n-type region will diffuse across the junction to the p-type side where they will recombine with some of the many holes in the p-type material. Similarly, holes will diffuse across the junction in the opposite direction and recombine, The recombination of free electrons and holes in the vicinity of the junction leaves a narrow region on either side of the junction that contains no mobile charge.

This narrow region which has been depleted of mibile charge is called the depletion layer. It extends into both the p-type and n-type regions as shown. Note that the diffusion of holes from the p-type side of the depletion layer leaves behind someuncovered fixed negative charges (the acceptor ions). Similarly, fixed positive charges (donor ions) are uncovered on the n-type side of the depletion layer. There is then a separation of charges: negative fixed charges on the p-type side of the depletion layer and positive fixed charges on the n-type side. This separation of charges causes an electric field to extend across the depletion layer. A potential difference must therefore exist across the depletion layer.

The uncovered charges give rise to a built-in potential of V_i_ volts. For a typical silicon p-n junction, V_i_ ≈ 0.6 to

0.7 volts. It varies with doping levels and temperature. The significance of this built-in potential is that it opposes the flow of holes and electrons across the junction. For this reason, the built-in potential is called a potential barrier or potential hill.

In practice, a p-n junction is formed within a single crystal rather than simply joining two pieces together. Electrical contacts on either side of the crystal enable connection to an external circuit. The resulting device is called a junction diode.

JUNCTION DIODE BEHAVIOUR

The most important property of a junction diode is its ability to pass an electric current in one direction only. If the diode is connected to a simple circuit consisting of a battery and a resistor, the battery can be connected in either of two ways.

When the p-type region of the p-n junction is connected to the positive terminal of the battery, current will flow. The diode is said to be under forward bias. However, when the battery terminals are reversed, the p-n junction almost completely blocks the current flow. This is called reverse bias. If the diode is not connected at all, it is said to be open-circuited and of course no current can flow through the diode.

FORWARD BIAS

The application of a forward bias voltage V to a junction diode reduces the built-in potential from V_i to $V_i - V$.

The reduction in the built-in potential is due to the applied voltage forcing more electrons into the n-type region and more holes into the p-type region, thus covering some of the fixed charges and narrowing the depletion layer. Since the total uncovered charge is reduced, the built-in potential must be lower.

Remembering that the built-in potential opposes the flow of majority carriers across the junction, a reduction in that potential makes it easier for holes in the p-type region to cross the junction and for electrons in the n-type region to cross the junction in the opposite direction. As the forward bias voltage is increased, the current through the junction becomes greater. When the applied voltage V approaches V_i, the potential hill is almost removed. There is then little opposition to the flow of carriers across the junction and a large current can flow through the diode.

The variation of diode current with voltage under forward bias is shown in the first quadrant of a typical junction diode current-voltage characteristic.

REVERSE BIAS

The application of a reverse voltage V_R extracts holes from the p-type region and free electrons from the n-type region and so uncovers more bound charges near the junction. The depletion layer therefore widens and the height of the potential hill is increased to $(V_i + V_R)$ volts. Majority carriers are thereby firther inhibited from crossing the junction. As the reverse voltage is increased, the current is reduced to almost zero.

However, a very small reverse current does flow. This reverse saturation current depends only on the thermal generation of holes and electrons near the junction, not on the height of the potential barrier. In practice, this reverse saturation current is quite small but it increases with increasing temperature.

JUNCTION BREAKDOWN

The large increase in reverse current evident in Figure is the result of junction breakdown. It occurs when the reverse voltage reaches a critical value.

THE DIODE EQUATION

A complete analysis of the abrupt p-n junction shows that that current I varies exponentially with applied voltage V. The exact relationship between current and voltage is given by the diode equation

$$I = I_S (e^{qV/kT} - 1)$$

where I_S is the saturation current, qis the electronic charge, k is Boltzmann's constant and T is temperature (\(deK). The diode equation applies for both forward and reverse bias. At extremes of high forward bias and large reverse voltages the behaviour of practical devices deviate from the above diode law. The diode equation is a very important description of diode behaviour and is the basis for the mathematical description of the behaviour of many other electronic devices which employ p-n junctions.

Exercise:

Derive an expression for the slope of the current-voltage characteristic of a forward-biased junction diode, and state how this slope can be related to the resistance of the diode.

JUNCTION TRANSISTORS

While a single PN junction is useful, it also has its limitations. It can either conduct currrent or not, but can't really control how much current it will conduct.

So what happens if we add a second junction? Can we play around with that idea and see where it goes?

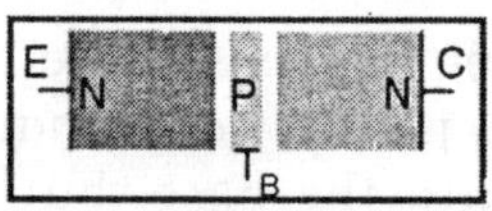

A basic two-junction semiconductor must necessarily have one type of region sandwiched between two of the other type. To the right is an example of a semiconductor device consisting of a narrow P-type region between two N-type regions. For reasons we will see shortly, the three regions are designated theemitter (E), base (B), and collector (C), respectively. In modern versions of this device, the emitter region is heavily doped with the appropriate impurity, while the base region is very lightly doped. The collector region has a moderate doping level so it will have a low internal resistance.

We have shown a device consisting of N, P, and N regions in order, but there is no reason not to build equivalent devices in P, N, and P order instead. In fact, it is often very useful to have both types of devices available.

With no electrical voltages applied, of course, this semiconductor will just sit there and do nothing. So let's move forward and see what happens when we apply bias voltages to the device.

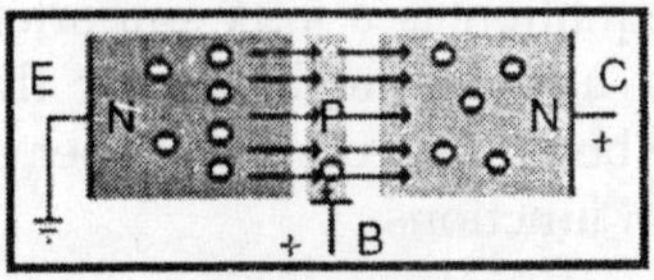

To the left we see the same semiconductor device as above, with a small forward bias applied to the emitter-base junction, and a larger reverse bias applied to the collector-base junction. As we will see, these are the normal operating conditions of this device.

Since we already know how a single-junction device, the diode, behaves, we would normally expect the base voltage to be about 0.65 to 0.7 volt positive with respect to the emitter, and to have electrons move from emitter to base, and leave the device at that point. With the collector junction reverse biased, we would expect no current to flow through that junction.

But a funny thing happens inside the base region. The forward bias on this junction does indeed attract electrons from the emitter into the base, but there the forward momentum of the electrons carries them across most of the base region and into the depletion region around the collector junction.

From there, the higher positive collector voltage attracts these electrons across the collector junction and into the collector region. (Remember that the electrons are minority current carriers within the P-type pase region, and can therefore cross the reverse-biased junction as a leakage current.)

A small amount of current does still leave the device through the base contact, but most of the current is diverted through the collector instead. In this way, the small base bias current controls the much larger collector current. If a small varying current is applied to the base along with the bias, the collector current will vary to a much greater degree. Thus, this device can not only be used to control a varying signal; it can amplify that signal as well.

Because of the way this device operates to transfer current (and its internal resistances) from the original conduction path to another, it's name is a combination of the words"transfer" and"resistor:" transistor.

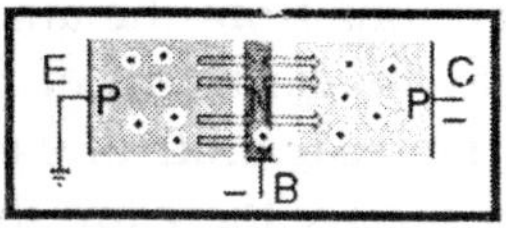

It's also quite possible to build a transistor with the region types reversed, as shown to the right. In this case, holes will be drawn from the emitter into the base region by the forward bias, and will then be pulled into the collector region by the higher negative bias.

Otherwise, this device works the same way and has the same general properties as the one described above. To distinguish between the two types of transistors, we refer to them by the order in which the different regions appear. Thus, this is a PNP transistor while the device described above is an NPN transistor.

ELECTROSTATIC FORCE

Electrostatics is associated with materials on which electrical charge moves only slowly (insulating materials) and with electrically isolated conductors. 'Insulation' and 'isolation' prevents easy migration of charge. So charges stay in place (are 'static'). It is the effects that the charges produce that are then important. The parameter of basic importance in electrostatics is 'charge'.

It is charge which gives rise to the electric fields generating forces that attract thin films and particles to surfaces and charge which gives rise to high voltages in low capacitance systems.

High voltages can cause sparks. If these have sufficient energy they may ignite flammable gases, cause shocks to personnel, damage semiconductor devices and upset the operation of microelectronic equipment.

Charge: Static electricity arises as the separation of positive and negative charges at the interface between two dissimilar surfaces.

If one or other of the surfaces prevent easy migration of charge, or the conductor on which they reside is isolated, then this charge is 'static' on the surface and remains available to influence the surroundings. 'Static' electricity can also arise on surfaces as trapped ions from the air.

Static charges may be electrons, or positive, or negative ions–but they are in the basic units of electronic charge 1.602 $\times 10^{-19}$ coulomb. On a surface there are some 10^{19} atomic sites per m^2–so if the charge of even quite a small fraction of the surface atoms is changed then quite large quantities of charge are easily involved.

FORCE BETWEEN CHARGES

The force between two charges, Q_1 and Q_2, is proportional to the quantities of charge, and inversely proportional to the square of the separation distance, *d*:

$$F = -Q_1 Q_2/(4\pi\varepsilon_0 d^2)$$

This is Coulomb's Law. The force *F* is in Newtons, the quantities of charge Q_1 and Q_2 are in Coulombs and the separation distance *d* is in meters. The constant ε_0 is called the 'permittivity of free space' and has a value 8.854×10^{-12}. Charges of the same polarity repel each other and of opposite polarity attract.

Examples of electrostatic forces are the attraction between thin layers of plastic film and of fine dust particles to surfaces.

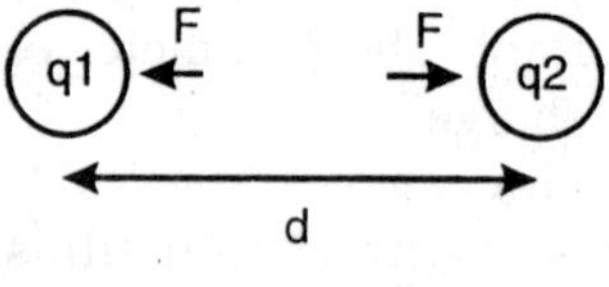

ELECTRIC FIELD

Around a charged body there is a force of attraction or repulsion for any other charges. This force, for a unit of charge, at any point is called the 'electric field', *E*, at that point. The direction of the field at any point depends on the direction of force on a positive charge there–and is hence a vector parameter. Diagrammatically it is useful to draw 'lines of force' to represent the electric field.

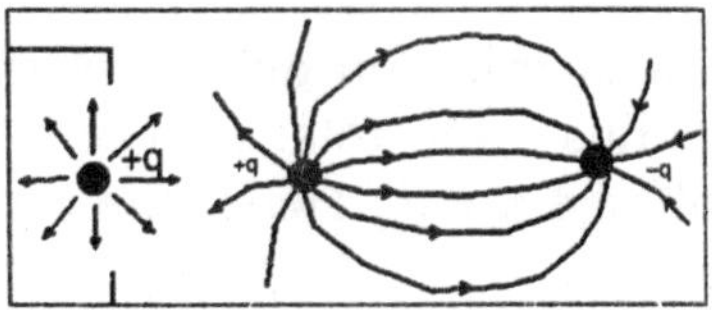

Fig. The Lines of Force the Stronger the Electric Field.

From Coulombs Law the electric field strength E at a distance d from a charge Q will be:

$$E = Q/(4\pi\varepsilon_0 d^2)$$

The constant of proportionality $1/(4\pi\varepsilon_0)$ is, for historical reasons, defined in the MKS system of units by the relation:

$$1/(4\pi\varepsilon_0) = c^2 \times 10^{-7}$$

The constant ε_0 hence has a value 8.854×10^{-12} coulomb²/(Newton m²) or coulomb/(volt m). This is the 'permittivity of free space' noted above.

POTENTIAL

The 'potential' at a point is defined as the amount of work needed to bring a unit charge from infinity to that point. The potential difference between two points is then the work done to move a unit charge between these two points. The work done does not depend on the route followed so the potential is a scalar quantity. The unit of potential is the 'volt'. The potential at a radial distance d from a single point charge Q is:

$$V = Q/(4\pi\varepsilon_0\, d)$$

The electric field relates to potential as:

$$E = -\nabla V$$

ELECTRIC FLUX }DENSITY

The electric field from an isolated point charge is radial. The field may conveniently be represented diagrammatically by lines of force. The 'flow' or flux of electric field across any spherical surface enclosing a point charge Q will then be Q/ε_0.

GAUSS' LAW

Gauss' Law states that the nett flux through any surface equals the nett charge enclosed within that surface.

$$\nabla . E = q/\varepsilon_0$$

For a uniform distribution density of charge ρ (C m^{-3}) in a sphere of radius r the total charge is $4/3\ \pi\ r^3\ \rho$. The outward flux will be $4\ \pi\ r^2\ E$. Hence the electric field at the sphere surface is:

$$E = \rho r/(3\varepsilon_0)$$

The maximum voltage at the centre of such a spherical distribution of charge is:

$$V = \rho r^2/(6\varepsilon_0)$$

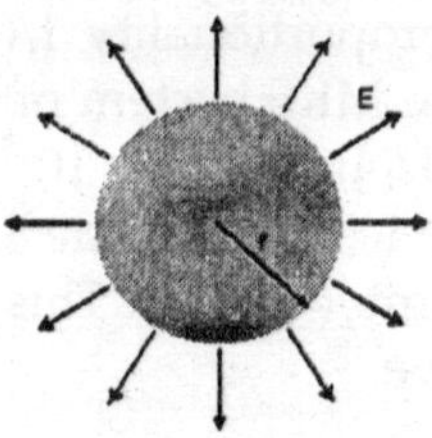

Similarly for a cylindrical distribution of charge density:

$$E = \rho r/(2\varepsilon_0) \text{ and } V = \rho r^2/(4\varepsilon_0)$$

INDUCED CHARGE

An electric field will have charge 'induced' on it with the density of charge proportional to the electric field. The charge induced on a conductor is of opposite polarity to that of the charge source. Opposite charges attract so there is an attraction between a charge and a conducting surface. For a plane conducting surface this 'image charge' is an equal distance behind the surface so calculation of the force involves a distance equal to twice the separation distance from the conducting surface. The electric field from each side of a sheet of uniform charge density σ (C m^{-2}) is:

$$E = \sigma/(2\varepsilon_0)$$

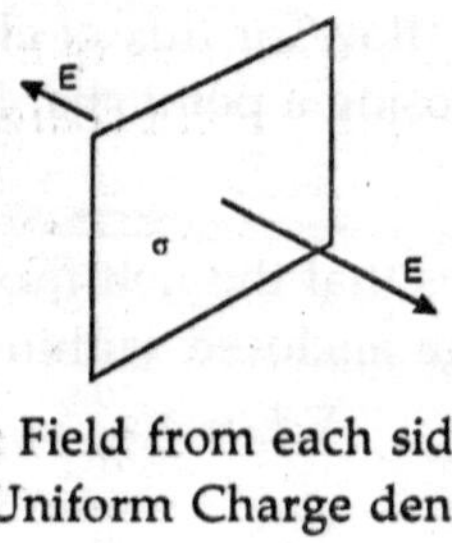

Fig. Electric Field from each side of a Sheet of Uniform Charge density

There can be no electric fields inside a conductor and the electric field outside is:

$$E = \sigma/\varepsilon_0$$

If a small conducting body (e.g. a small sphere) on an insulating handle is touched to a conducting body charge will be shared and the probe will acquire a charge dependent of the local electric field–related to the local surface charge density.

If the amount of charge can be tested then we have a means to check if the conducting body was charged and to explore the density of charge over that body.

A 'gold leaf electroscope' indicates charge by repulsion between a fine gold leaf from it's nearby conducting support and gives an angle of the leaf indicating the level of charge. Sharing charge from a conducting probe with a gold leaf electroscope provides a good classical method to indicate and measure charge.

Such observations show that the electric field (and charge density) on a conducting body is inversely proportional to the local radius of curvature.

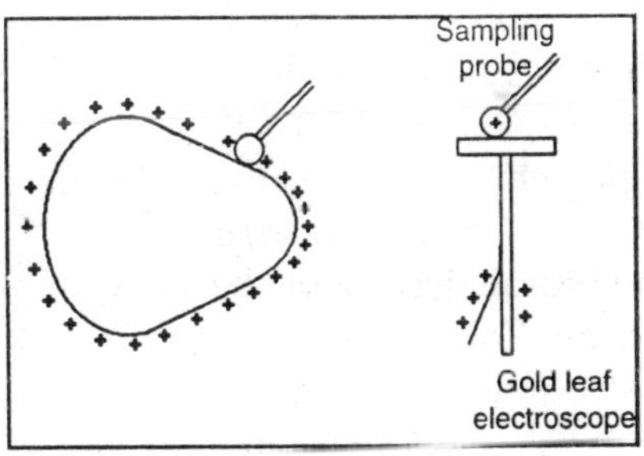

CAPACITANCE

For two conducting plates of area A separated by a distance d(m) with a voltage difference between them V the electric field between the plates is:

$$E = V/d$$

and the quantity of charge on each of the plates:

$$Q = A\,\sigma = \varepsilon_0\,AE = \varepsilon_0\,AV/d$$

The proportionality between the quantity of charge Q and the voltage V is called the 'capacitance' C of the system. The unit of capacitance is the 'farad' with units of coulomb/volt. For the pair of parallel plates:

$$C = Q/V = \varepsilon_0 A/d$$

This relation is not quite accurate because the electric field is not uniform near the edges of the plates and extends outwards from them a bit.

For an isolated sphere of radius a (m) the capacitance is $4\pi\varepsilon_0 a$– for a radius of 10mm this is about 1 picofarad. The capacitance for a number of simple geometric arrangements (for instance concentric spheres, concentric cylinders, parallel rods and rods near plane conductors) can be caculated analytically.

These can provide useful guidance for estimating capacitance in practical situations. Alternatively capacitance valus may be calculated by computer modelling calculations or, with suitable care, measured. The electrostatic energy U (joule) of a capacitor of C (farads) is:

$$U = 0.5\ CV^2$$

Poisson Equation:

$$\nabla^2 V = -\rho/\varepsilon_0$$

where ñ is the volume density of charge (C m^{-3}). If no charges are present then Poisson's equation becomes Laplace's equation:

$$\nabla^2 V = 0$$

There are a number of geometric forms for which the above equations can be solved analytically. Two and three dimensional finite element and finite difference computer modelling programs are available to find potential and electric field distributions within practical geometric arrangements.

DIELECTRICS

If a block of conductor were placed between the capacitor

plates the field in the remaining gap would be increased. A similar effect occurs with insulating materials, but because charges cannot move through or over the surface of insulators the material becomes 'polarised'. The atoms or molecules distort slightly under the electrical stress, polarise, and while these effects cancel out within the insulator they appear as surface charges on either side of the block of material. If an insulator is placed between the plates of a parallel plate capacitor the capacitance is increased.

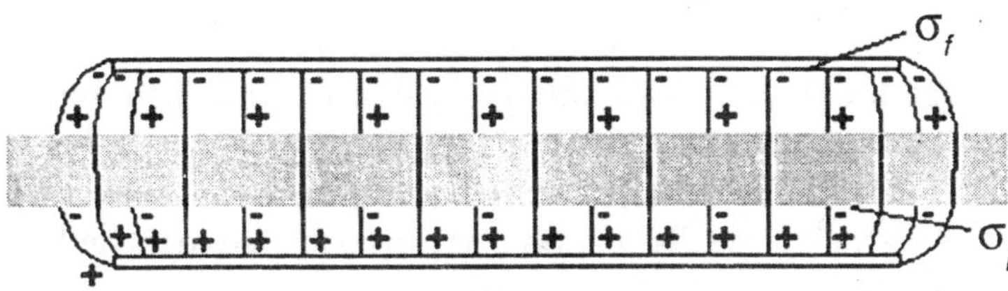

Applying Gauss's Law to the surface S the electric field in the insulator is:

$$E = (\sigma_f - \sigma_p)/\varepsilon_0$$

where σ_f and σ_p are the charge densities on the free surface of the capacitor plates and the surface of the insulator. The polarisation charges are proportional to the electric field and this is usually written:

$$P = X\,\varepsilon_0 E = \sigma_p$$

$$E = \sigma_f/\varepsilon_0(1 + X))$$

where X is the 'electric susceptibility' of the insulator/dielectric. This shows the reduction of the electric field within the insulator compared to that just outside. The total charge on the capacitor plates is $\sigma_f A$ so the capacitance by $Q = CV$ becomes for a capacitor fully filled by an insulator:

$$C = \sigma_f A/V = AE\,(0(1 + X))/(E\,d)$$

$$C = \varepsilon_0 A\,(1 + X)/d$$

The increase in capacitance for this fully filled capacitor is the factor (1 + X) and this is usually called the 'Permittivity' of the material –*k*.

CHARGE MIGRATION AND DISSIPATION

Insulating materials have very few free electrical charges

so only very small currents flow under the influence of an electric field. In more usual terms the volume 'resistivity' may be from 10^{12} to 10^{16} ohm m and the surface resistivity over 10^{10} ohms per square. Not only are there few charge carriers but their number and the velocity of their movement may depend in a non-linear way upon the electric field.

In this situation 'resistivity' is not a very useful parameter for characterising materials. Of more direct practical importance in the control of static is the timescale for static charge dissipation. If the timescale for charge dissipation is less than the timescale for charge generation then no significant charge levels or surface voltages will arise and there will be no static problems.

For uniform materials, such as liquids, it is expected that the time for charge dissipation $\tau(\sigma)$ can be related to the volume resistivity ρ (ohms) and permittivity k as:

$$\tau = k\,\varepsilon_0\rho$$

CHARGE SEPARATION MECHANISMS

Contact between surfaces results in transfer of charge as electrons to match up energy levels of neighbouring atoms around the points of contact. The flow of charge continues until inhibited by the potential difference of the 'contact potential' of the surfaces. On separation the surfaces may retain the transferred charge.

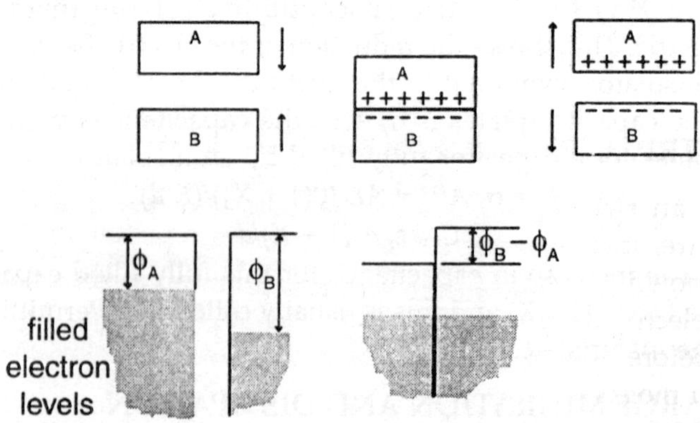

Charge transfer depends on differences in 'contact

potential' and materials can be listed as a 'triboelectric series'. Charge separation depends on surface conditions. It is strongly affected by contamination and by the speed and pressure of rubbing actions.

Only in a few instances, for example in photocopiers and in some paint spray situation, are triboelectric effects used to control charging in a defined way. Charge separation also occurs when a liquid meets a conducting surface. Ions formed by dissociation in the fluid or from trace impurities will adsorb at the boundary surface.

This creates the so-called 'double layer' at the boundary. An appreciable charge can be adsorbed to the surface–which is matched by charge in the fluid. With relatively insulating fluids the charge in the volume can be swept away by fluid flow as a 'streaming current' which can be in the range 10^{-10} to 10^{-7} A. A double layer also occurs when water breaks away from a surface–and this is responsible for the formation of the highly charged mists that can arise when high pressure water jets hit surfaces.

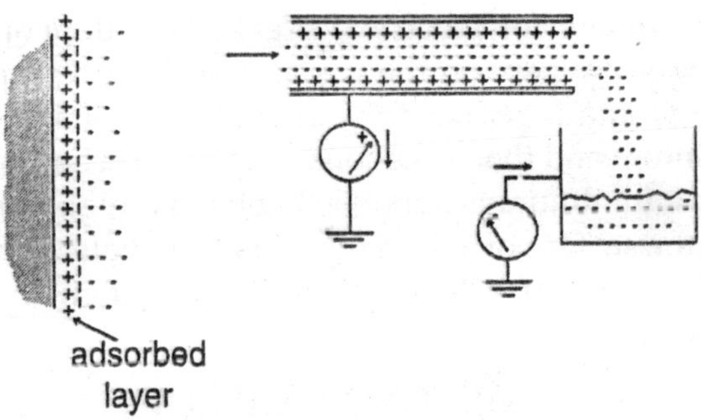

ELECTRICAL BREAKDOWN

If an electric field, in air at normal temperature and pressure, increases above about 3 MV m^{-1} then 'electrical breakdown' occurs. A loose electron or ion accelerates in the local electric field and if it picks up enough energy from the field before it hits another air molecule it will cause ionisation of that molecule.

This can lead to further ionisation and rapid growth of

the number of ions present. This multiplication of ionisation can destroy the insulating properties of the air and lead to passage of a high current accompanied by the emission of heat and light and sound–this is a spark discharge.

If the electric field exceeds the critical value for ion multiplication in just a localised region, for example near a sharp point, while the average electric field over the gap to nearby conductors is quite low, then a localised discharge takes place with a fairly low current (up to perhaps a mA) between the electrodes. This is a 'corona' discharge. If the electric field exceeds the breakdown strength of air over several mm, for example with an electrode of several mm radius of curvature, then the discharge involves more current and has a condensed luminous channel which does not reach to another electrode. This is a 'brush discharge'.

If the breakdown strength of air is exceeded near an electrode and also the average electric field over the gap to the other electrode exceeds about 500 kV m^{-1} then breakdown is likely to occur as a spark discharge between the electrodes.

The voltage for electrical breakdown depends on the geometry of the gap, on the gas and on the pressure. As the pressure or gap is reduced the breakdown voltage decreases to a minimum and then rises-following the Paschen Law curve. For air the minimum is about 300 volts. If a high electric field arises at the surface of a fluid the surface will distort and break up. This occurs at a lower electric field than that for air breakdown.

COULOMB'S LAW OF ELECTROSTATIC FORCES

CHARGE INTERACTIONS BY COULOMB'S LAW

Electrical interactions between charges govern much of physics, chemistry, and biology. They are the basis for chemical bonding, weak and strong. Salts dissolve in water to form the solutions of charged ions that cover three-quarters of the Earth's surface. Salt water forms the working fluid of living cells. pH and salts regulate the associations of proteins,

DNA, cells, and colloids, and the conformations of biopolymers.

Nervous systems would not function without ion fluxes. Electrostatic interactions are also important in batteries, corrosion, and electroplating. Charge interactions obey Coulomb's law. When more than two charged particles interact, the energies are sums of coulombic interactions.

With these tools you can determine the electrostatic force exerted by one charged object on another, as when an ion interacts with a protein, DNA molecule, or membrane, or when a charged polymer changes conformation. Coulomb's law was discovered in careful experiments by H Cavendish, J Priestley, and CA Coulomb (1736–1806) on macroscopic objects such as magnets, glass rods, charged spheres, and silk cloths. But Coulomb's law applies to a wide range of size scales, including atoms, molecules and biological cells. It states that the interaction energy *u(r)* between two charges in a vacuum is,

$$u(r) = e\frac{q_1 q_2}{r},$$

where q_1 and q_2 are the magnitudes of the two charges, r is the distance separating them, and $e = 1/4\pi\varepsilon_0$ is a proportionality constant.

CHARGE INTERACTIONS BY LONG-RANGED

Uncharged atoms interact with each other through *short-ranged* interactions. That is, attractions between uncharged atoms are felt only when the two atoms are so close to each other that they nearly contact.

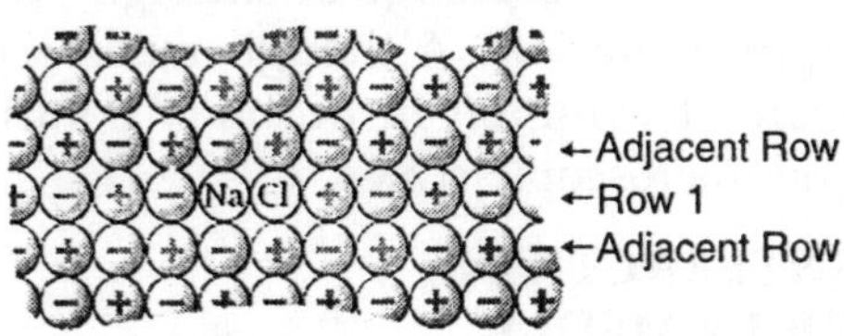

Fig. Crystalline Sodium Chloride Packs in a Cubic Array of Alternating Sodium and Chloride Ions

Short-ranged interactions typically diminish with distance as r^{-6} it suffices to note that charge interactions are very different: they are long-ranged. Coulombic interactions diminish with distance as r^{-1}. Counting lattice contacts is no longer sufficient. The difference between short-ranged and long-ranged interactions is much more profound than it may seem. The mathematics and physics of longranged and short-ranged interactions are very different.

A particle that interacts through short-ranged interactions feels only its nearest neighbors, so system energies can be computed by counting the nearest-neighbour contacts. But when interactions involve charges, more distant neighbors contribute to the energies, so our methods of summing energies must become more sophisticated.

Electrostatic interactions are strong as well as long-ranged. The electrostatic repulsion is about 1036 times stronger than the gravitational attraction between two protons. Strong, short-ranged nuclear forces are required to hold the protons together against their electrostatic repulsions in atomic nuclei.

When charge builds up on a macroscopic scale, violent events may result, such as lightning and thunder storms, and explosions in oil tankers. Beçause charge interactions are so strong, matter does not sustain charge separations on macroscopic scales: macroscopic matter is neutral.

The transfer of less than a thousand charges is enough to cause the static electricity that you observe when you rub your shoes on a carpet. In solution, too, imbalances are usually less than thousands of charges, but because this is one part in 1021, it means that you can assume that solutions have charge neutrality—no net charge of one sign—to a very high degree of approximation. Although macroscopic charge separations aren't common in ordinary solutions, charge separations on microscopic scales.

CHARGE INTERACTIONS BY WEAKER IN MEDIA

Charges interact more weakly when they are in liquids

than when they are in a vacuum. To varying degrees, liquids can be *polarized*. This means that if two fixed charges, say a negative charge A and a positive charge B, are separated by a distance r in a liquid, then the intervening molecules of the liquid tend to reorient their charges.

The positive charges in the medium orient toward A, and the negative charges toward B, to shield and weaken the interaction between A and B. This weakening of the coulombic interactions between A and B is described by the *dielectric constant* of the medium. Media that can be readily polarized shield charge interactions strongly: they have high dielectric constants.

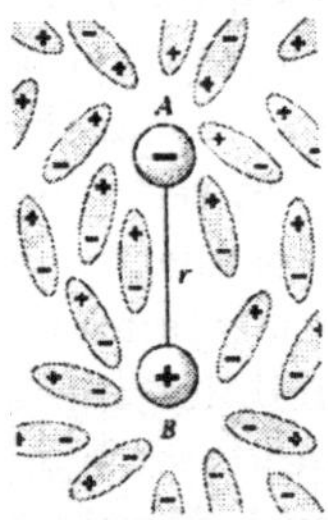

Fig. Fixed Charges A and B Cause Dipoles in the Surrounding Liquid to Orient Partly, Which Causes some *Shielding*, a Reduction of the Interaction Between A and B.

The medium as a *polarizable isotropic continuum*. In treating a medium as a continuum, we neglect its atomic structure and focus on its larger-scale properties. In treating a medium as *isotropic*, we assume that its polarizability is the same in all directions. By treating a medium as *polarizable*, we assume that the charge redistributes in response to an electric field, even if the medium is neutral overall.

The more polarizable the medium, the greater is the reduction of an electrostatic field across it. The reduction factor is the *dielectric constant D*, a scalar dimensionless quantity. When charges q_1 and q_2 are separated by a distance r in a medium of dielectric constant D, their coulombic interaction energy is,

$$u(r) = \frac{eq_1q_2}{Dr}.$$

Coulombic interactions are traditionally described as an energy *u(r)*, but when dielectric constants are involved, *u(r)* is a free energy, because the dielectric constant is temperature dependent. The polarizability of a medium arises from several factors. First, if the medium is composed of molecules that have a *permanent dipole moment,* a positive charge at one end and a negative charge at the other, then applying an electric field tends to orient the dipoles of the medium in a direction that opposes the field.

Second, the polarizability of a medium can arise from the polarizabilities of the atoms or molecules composing it, even if they have no permanent dipole moment. Atoms or molecules that lack permanent dipoles have electronic *polarizability,* a tendency of nuclear or electronic charge distributions to shift slightly within the atom, in response to an electric field. The electronic polarizabilities of hydrocarbons and other nonpolar substances are the main contributors to their dielectric constants ($D \approx 2$). Third, polarizabilities can also arise from networks of hydrogen bonds in liquids such as water. If there are many alternative hydrogen-bond donors and acceptors, a shift in the pattern of hydrogen bonding is not energetically costly, so the networks of hydrogen bonds in hydrogen-bonded liquids are labile when electric fields are applied.

THE BJERRUM LENGTH

To show the effect of the polarizability of the medium, let's first compute a useful quantity called the *Bjerrum length –B,* defined as the charge separation at which the coulomb energy between ions *u(r)* just equals the thermal energy *RT.*

Substitute ℓ_B for r and RT for u in $u(r) = eq_1q_2/Dr$ and solve for ℓ_B forsingle charges $q_1 = q_2 = e$:

$$\ell_B = \frac{ee^2N}{DRT}.$$

For a temperature $T = 298.15$, in a vacuum or air ($D = 1$),

$$\ell_B = \frac{1.386 \times 10^{-4}\,\text{J m mol}^{-1}}{(8.314\,\text{J K}^{-1}\text{mol}^{-1} \times 298.15K)} = 560Å.$$

560 Å is a relatively long distance, many times the diameter of an atom. When two charges are much closer together than this distance, they feel a strong repulsive or attractive interaction, much larger than the thermal energy *RT*. But when two charges are much further apart than this distance, their interactions are weaker than *RT*, and they are more strongly directed by Brownian motion than by their coulombic interactions.

Dividing $u(r) = eq_1q_2/Dr$ by $(RT = eq_1q_2/D\ell_B)$ gives,

$$\frac{u}{RT} = \frac{\ell_B}{r}.$$

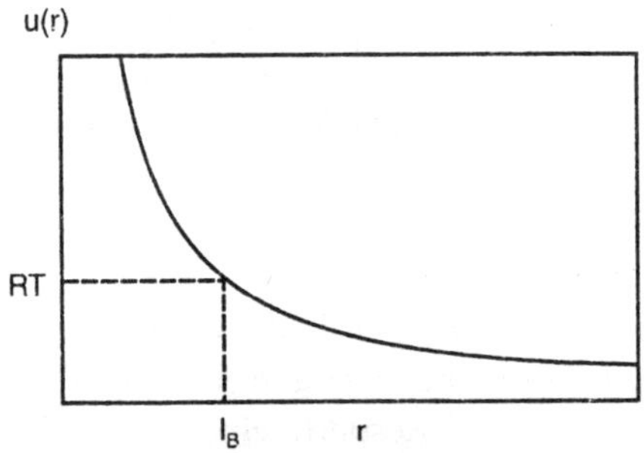

Fig. The Bjerrum Length *B* is Defined as the Distance *r* at Which two Unit Charges Interact with an Energy *u(r)* Equal to *RT*.

So far, our calculations for ionic interactions have involved simple arrangements of charges. When several charges are involved, or when charges have complex spatial arrangements, more sophisticated methods are needed to treat them. We now develop some useful tools that lead to Gauss's law and Poisson's equation.

ELECTROSTATIC FORCES ADD LIKE VECTORS

Vectors and Coulomb's Law

Coulomb's law may be expressed either in terms of energy $u(r) = eq_1q_2/Dr$, or force $f = -(\partial u/\partial r) = eq_1q_2/Dr^2$.

The difference is that energies are scalar quantities, which add simply, while forces are vectors that add in component form. Either way, an important fundamental law is the superposition principle: electrostatic energies and forces are both additive. The total electrostatic force on a particle is the vector sum of the electrostatic forces from all other particles.

$$f = \frac{eq_A q_B}{Dr^2} \cdot \frac{r}{r},$$

Where *r*/*r* indicates a vector pointing in the direction of vector *r*, along the line connecting the particles, but with unit length. *f* is the force acting in the same direction.

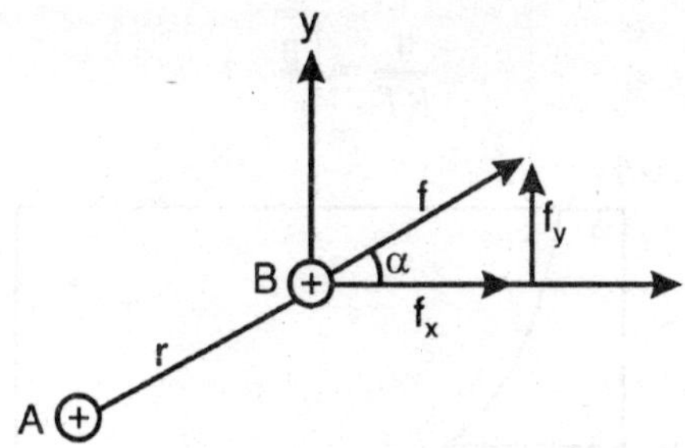

Fig. The Electrostatic Force *f* of Fixed Charge *A* on test Charge *B* is a Vector with Components *fx* and *fy* Along the *x*- and *y*-axes, Respectively.

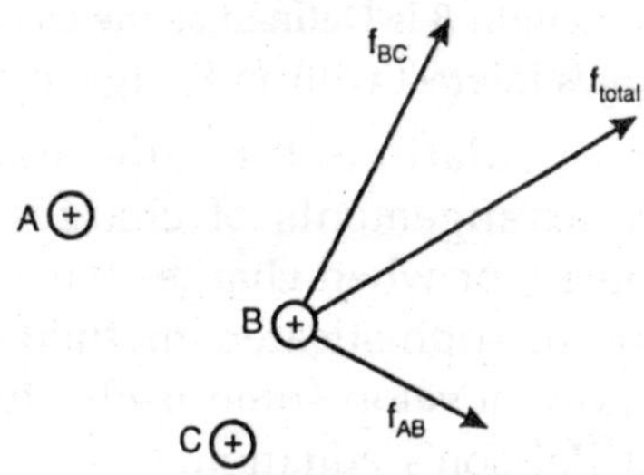

Fig. The Force on the Test Charge *B* is the Vector Sum from the two Fixed Charges *A* and *C*.

The component of the force *f* in the *x*-direction is described by using the component of the vector *r* between the charges, *rx* = *r* cos α, where α is the angle between *f* and the *x*-axis. The *x*-direction component of the force is,

$$f_x = \left(\frac{q_A q_B}{4\pi\varepsilon_0 Dr^2}\right)\cos\alpha.$$

The y-direction component is:

$$f_y = \left(\frac{q_A q_B}{4\pi\varepsilon_0 D r^2}\right)\sin\alpha.$$

ELECTRIC FIELD

In physics, the space surrounding an electric charge or in the presence of a time-varying magnetic field has a property called an electric field (that can also be equated to electric flux density). This electric field exerts a force on other electrically charged objects. The concept of electric field was introduced by Michael Faraday.

The electric field is a vector field with SI units of newtons per coulomb ($N\ C^{-1}$) or, equivalently, volts per meter ($V\ m^{-1}$). The strength of the field at a given point is defined as the force that would be exerted on a positive test charge of +1 coulomb placed at that point; the direction of the field is given by the direction of that force. Electric fields contain electrical energy with energy density proportional to the square of the field intensity. The electric field is to charge as gravitational acceleration is to mass and force density is to volume.

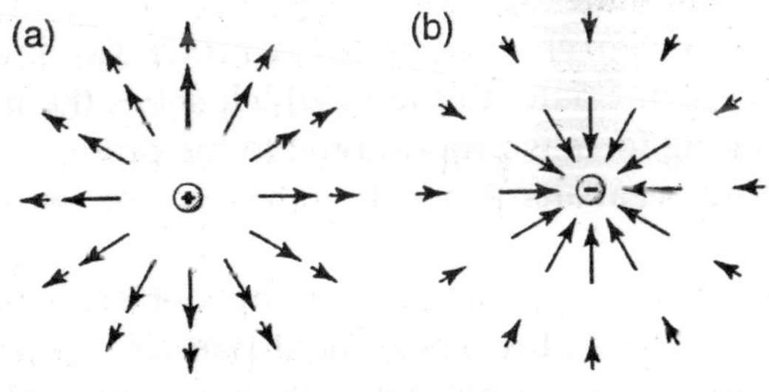

Fig. The Electric Field is indicated by Vectors Pointing.

(a) Away from a Positive Charge

(b) Toward a Negative Charge.

The Field Represents the Force Per Unit Charge that acts on a Positive Test Charge Put at Each Particular Position in Space. A moving charge has not just an electric field but also a magnetic field, and in general the electric and magnetic fields are not completely separate phenomena; what one observer perceives as an electric field, another observer in a different frame of reference perceives as a mixture of electric and magnetic fields.

For this reason, one speaks of "electromagnetism" or "electromagnetic fields." In quantum mechanics, disturbances in the electromagnetic fields are called photons, and the energy of photons is quantized. A stationary charged particle in an electric field experiences a force proportional to its charge given by the equation,

$$F = q\left[-\nabla\phi - \frac{\partial A}{\partial t}\right]$$

where the magnetic flux density is given by,

$$B = \nabla \times A$$

and where $-\nabla\phi$ is the Coulomb force.

Electric charge is a characteristic of some subatomic particles, and is quantized when expressed as a multiple of the so-called elementary charge e. Electrons by convention have a charge of –1, while protons have the opposite charge of +1. Quarks have a fractional charge of –1/3 or +2/3. The antiparticle equivalents of these have the opposite charge. There are other charged particles.

In general, same-sign charged particles repel one another, while different-sign charged particles attract. This is expressed quantitatively in Coulomb's law, which states the magnitude of the repelling force is proportional to the product of the two charges, and weakens proportionately to the square of the distance.

The electric charge of a macroscopic object is the sum of the electric charges of its constituent particles. Often, the net electric charge is zero, since naturally the number of electrons in every atom is equal to the number of the protons, so their charges cancel out. Situations in which the net charge is non-zero are often referred to as static electricity.

Even when the net charge is zero, it can be distributed non-uniformly (e.g., due to an external electric field), and then the material is said to be polarized, and the charge related to the polarization is known as bound charge (while the excess charge brought from outside is called free charge).

An ordered motion of charged particles in a particular direction (in metals, these are the electrons) is known as electric current. The discrete nature of electric charge was proposed by Michael Faraday in his electrolysis experiments, then directly demonstrated by Robert Millikan in his oil-drop experiment.

The SI unit for quantity of electricity or electric charge is the coulomb, which represents approximately 1.60×10^{19} elementary charges (the charge on a single electron or proton). The coulomb is defined as the quantity of charge that has passed through the cross-section of an electrical conductor carrying one ampere within one second. The symbol Q is often used to denote a quantity of electricity or charge.

The quantity of electric charge can be directly measured with an electrometer, or indirectly measured with a ballistic galvanometer. Formally, a measure of charge should be a multiple of the elementary charge e (charge is quantized), but since it is an average, macroscopic quantity, many orders of magnitude larger than a single elementary charge, it can effectively take on any real value.

In some contexts it is meaningful to speak of fractions of a charge; e.g. in the charging of a capacitor. If the charged particle can be considered a point charge, the electric field is defined as the force it experiences per unit charge:

$$E = \frac{F}{q}$$

where as:

F is the electric force experienced by the particle;

q is its charge;

E is the electric field wherein the particle is located;

Taken literally, this equation only defines the electric field at the places where there are stationary charges present to experience it. The force exerted by another charge q will alter the source distribution, which means the electric field in the presence of q differs from itself in the absence of q. However, the electric field of a given source distribution remains defined in the absence of any charges with which to interact.

This is achieved by measuring the force exerted on successively smaller *test charges* placed in the vicinity of the source distribution. By this process, the electric field created by a given source distribution is defined as the limit as the test charge approaches zero of the force per unit charge exerted there upon.

$$E = \lim_{q \to 0} \frac{F}{q}$$

This allows the electric field to be dependent on the source distribution alone. As is clear from the definition, the direction of the electric field is the same as the direction of the force it would exert on a positively-charged particle, and opposite the direction of the force on a negatively-charged particle. Since like charges repel and opposites attract the electric field tends to point away from positive charges and towards negative charges

Electric Fields have Fluxes

The electric field flux has the important general property that it is independent of the size and shape of any balloon that fully surrounds all the charges, as we will show.

This generality leads to some important methods for simplifying the problem of computing complex electrostatic fields. Positive and negative charges are the sources and sinks of electrical fields.

Consider some constellation of positive and negative charges fixed at certain positions in space. Just as we did for fluids we invent an imaginary balloon that we place around the charges. We want to determine the flux of the electric field through the imaginary balloon surface.

The electric field flux Φ is defined as the integral of D times the electric field E over all the surface area elements ds of the balloon, where D is the dielectric constant. Because E is proportional to $1/D$ in a uniform medium the flux Φ is independent of D.

$$\Phi = \int_{\text{surface}} DE.ds,$$

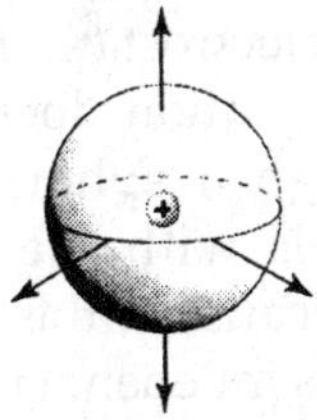

Fig. The Vector Electric Field E is Normal to a Sphere Surrounding a Point Charge located at the Centre of the Sphere

Equation is general and applies to any constellation of fixed charges, and to any shape and position of an imaginary balloon around them. Many problems of interest have a special simplicity or symmetry that allows you to simplify Equation.

For example, suppose that the charge constellation is just a single point charge q and that the imaginary balloon is just a sphere of radius r centered on the point charge. We are interested in computing the electric field flux from the point charge through the imaginary spherical surface.

The electrostatic field of a point charge is spherically symmetrical—the magnitude of the field E is the same through the north pole as it is through the south pole or through the equator, or through any other patch of surface of a given area. So you can drop the vector notation and express the field as $E(r)$, which depends only on r. Because E doesn't differ from one surface patch to another, the integral in Equation becomes a simple product. That is, E is parallel to ds everywhere, so $E \cdot ds = Eds$ and the flux integral becomes:

$$\Phi = DE \int ds = DE(r) \times \text{(total area of sphere)}.$$

Use $E(r) = q/(4\pi\varepsilon_0 Dr^2)$ from Equation and $4\pi r^2$ for the surface area of a sphere, to get

$$\Phi = DE(r)(4\pi r^2)$$

$$= \left(\frac{q}{4\pi\varepsilon_0 r^2}\right)(4\pi r^2)$$

$$= \frac{q}{\varepsilon_0},$$

where q is the charge at the centre of the sphere.

Two profound implications of Equation lead to a general method for calculating electrostatic forces and fields. First, the right-hand side of Equation does not depend on r. For the electric flux through a spherical container, the r^{-2} dependence of E cancels with the r^2 dependence of the spherical surface area. Because of this cancelation, the electric field flux out of a sphere is independent of its radius. It doesn't matter how large you choose your imaginary balloon to be, the flux is the same through spherical balloons of any radius.

POISSON'S EQUATION

In mathematics, Poisson's equation is a partial differential equation with broad utility in electrostatics, mechanical engineering and theoretical physics. It is named after the French mathematician, geometer and physicist Siméon-Denis Poisson. The Poisson equation is

$$\Delta\varphi = f$$

where Δ is the Laplace operator, and f and ö are real or complex-valued functions on a manifold. When the manifold is Euclidean space, the Laplace operator is often denoted as ∇^2 and so Poisson's equation is frequently written as:

$$\nabla^2\varphi = f$$

In three-dimensional Cartesian coordinates, it takes the form:

$$\left(\frac{\partial^2}{\partial x^2}+\frac{\partial^2}{\partial y^2}+\frac{\partial^2}{\partial z^2}\right)\varphi(x,y,z) = f(x,y,z).$$

For vanishing f, this equation becomes Laplace's equation:

$$\Delta\varphi = 0$$

The Poisson equation may be solved using a Green's function; a general exposition of the Green's function for the Poisson equation is given in the article on the screened Poisson equation. There are various methods for numerical solution. The relaxation method, an iterative algorithm, is one example.

ELECTROSTATICS

One of the principal cornerstones of electrostatics is the

posing and solving of problems that are described by the Poisson equation. Finding φ for some given f is an important practical problem, since this is the usual way to find the electric potential for a given charge distribution. The derivation of Poisson's equation in electrostatics follows. SI units are used and Euclidean space is assumed. Starting with Gauss' law for electricity in a differential control volume, we have:

$$\nabla . D = \rho$$

∇ means to take the divergence.

D is the electric displacement field.

ρ is the charge density.

Assuming the medium is linear, isotropic, and homogeneous, then:

$$D = \varepsilon E = \varepsilon_r \varepsilon_0 E$$

ε is the permittivity of the medium.

E is the electric field.

ε_0 is the vacuum permittivity.

ε_r is the relative permittivity of the medium.

By substitution and division, we have:

$$\nabla . E = \frac{\rho}{\varepsilon_r \varepsilon_0}$$

In the absence of a changing magnetic field, B, Faraday's law of induction gives:

$$\nabla \times E = -\frac{\partial B}{\partial t} = 0$$

$\times$ is the cross product.

t is time.

Since the curl of the electric field is zero, it is defined by a scalar electric potential field, φ

$$E = -\nabla \varphi$$

Eliminating E by substitution, we have a form of the Poisson equation:

$$\nabla . \nabla \varphi = \nabla^2 \varphi = -\frac{\rho}{\varepsilon_r \varepsilon_0}$$

Solving Poisson's equation for the potential requires knowing the charge density distribution. If the charge density is zero, then Laplace's equation results. If the charge density follows a Boltzmann distribution, then the Poisson-Boltzmann equation results.

The Poisson-Boltzmann equation plays a role in the development of the Debye-Hückel theory of dilute electrolyte solutions. The magnetic field not varying in time, the same Poisson equation arises even if it does vary in time, as long as the Coulomb gauge is used.

However, in this more general context, computing φ is no longer sufficient to calculate E, since the latter also depends on the magnetic vector potential, which must be independently computed.

POTENTIAL OF A GAUSSIAN CHARGE DENSITY

If there is a spherically symmetric Gaussian charge density $\rho(r)$:

$$\rho(r) = \frac{Q}{\sigma^3 \sqrt{2\pi}^3} e^{-r^2/(2\sigma^2)},$$

where Q is the total charge, then the solution $\varphi(r)$ of Poisson's equation with $\varepsilon_r = 1$,

$$\nabla^2 \varphi = -\frac{\rho}{\varepsilon_0},$$

is given by

$$\varphi(r) = \frac{1}{4\pi\varepsilon_0} \frac{Q}{r} erf\left(\frac{r}{\sqrt{2}\sigma}\right).$$

where erf(x) is the error function. This solution can be checked explicitly by a careful manual evaluation of $\nabla^2 \varphi$. Note that, for r much greater than σ, erf(x) approaches unity and the potential $\phi(r)$ approaches the point charge potential,

$$\frac{1}{4\pi\varepsilon_0} \frac{Q}{r}$$

as one would expect. The erf function approaches 1 extremely fast as its argument increase; in practice for $r > 3\sigma$ the relative error is smaller than 1/1000.

LAPLACE'S EQUATION

Laplace's equation is a partial differential equation named after Pierre-Simon Laplace who first studied its properties. The solutions of Laplace's equation are important in many fields of science, notably the fields of electromagnetism, astronomy, and fluid dynamics, because they describe the behaviour of electric, gravitational, and fluid potentials.

The general theory of solutions to Laplace's equation is known as potential theory. In the study of heat conduction, the Laplace equation is the steady-state heat equation. In three dimensions, the problem is to find twice-differentiable real-valued functions, φ of real variables, x, y, and z, such that:

$$\frac{\partial^2 \varphi}{\partial x^2} + \frac{\partial^2 \varphi}{\partial y^2} + \frac{\partial^2 \varphi}{\partial z^2} = 0.$$

This is often written as:

$$\nabla^2 \varphi = 0$$

or

$$\text{div grad } \varphi = 0,$$

where div is the divergence, and grad is the gradient, or

$$\Delta \varphi = 0,$$

where Δ is the Laplace operator.

Solutions of Laplace's equation are called harmonic functions.

If the right-hand side is specified as a given function, $f(x, y, z)$, i.e.

$$\Delta \varphi = f$$

then the equation is called "Poisson's equation." Laplace's equation and Poisson's equation are the simplest examples of elliptic partial differential equations. The partial differential operator, ∇^2, or Δ, (which may be defined in any number of dimensions) is called the Laplace operator, or just the Laplacian.

BOUNDARY CONDITIONS

The Dirichlet problem for Laplace's equation consists of finding a solution φ on some domain *D* such that φ on the boundary of *D* is equal to some given function. Since the Laplace operator appears in the heat equation, one physical interpretation of this problem is as follows: fix the temperature on the boundary of the domain and wait until the temperature in the interior doesn't change anymore; the temperature distribution in the interior will then be given by the solution to the corresponding Dirichlet problem.

The Neumann boundary conditions for Laplace's equation specify not the function φ itself on the boundary of *D*, but its normal derivative. Physically, this corresponds to the construction of a potential for a vector field whose effect is known at the boundary of *D* alone.

Solutions of Laplace's equation are called harmonic functions; they are all analytic within the domain where the equation is satisfied.

If any two functions are solutions to Laplace's equation (or any linear homogenous differential equation), their sum (or any linear combination) is also a solution. This property, called the principle of superposition, is very useful, e.g., solutions to complex problems can be constructed by summing simple solutions.

LAPLACE EQUATION IN TWO DIMENSIONS

The Laplace equation in two independent variables has the form

$$\varphi_{xx} + \varphi_{yy} = 0.$$

Analytic Functions

The real and imaginary parts of a complex analytic function both satisfy the Laplace equation. That is, if $z = x + iy$, and if

$$f(z) = u(x, y) + iv(x, y),$$

then the necessary condition that $f(z)$ be analytic is that the Cauchy-Riemann equations be satisfied:

$$u_x = v_y,\ v_x = -u_y.$$

It follows that:

$$u_{yy} = (-v_x)_y = -(v_y)_x = -(u_x)_x.$$

Therefore u satisfies the Laplace equation. A similar calculation shows that v also satisfies the Laplace equation.

Conversely, given a harmonic function, it is the real part of an analytic function, $f(z)$ (at least locally).

If a trial form is,

$$f(x) = \varphi(x, y) + i\psi(x, y),$$

then the Cauchy-Riemann equations will be satisfied if we set

$$\psi_x = \psi_y, \ \psi_y = \varphi x.$$

This relation does not determine ø, but only its increments:

$$d\psi = -\varphi y dx + \varphi_x dy.$$

The Laplace equation for φ implies that the integrability condition for ψ is satisfied:

$$\psi_{xy} = \psi_{yx}.$$

and thus ψ may be defined by a line integral. The integrability condition and Stokes' theorem implies that the value of the line integral connecting two points is independent of the path. The resulting pair of solutions of the Laplace equation are called conjugate harmonic functions. This construction is only valid locally, or provided that the path does not loop around a singularity. For example, if r and q are polar coordinates and,

$$\varphi = \log r,$$

then a corresponding analytic function is

$$f(x) = \log z = \log r + i\theta.$$

However, the angle 0 is single-valued only in a region that does not enclose the origin. The close connection between the Laplace equation and analytic functions implies that any solution of the Laplace equation has derivatives of all orders, and can be expanded in a power series, at least inside a circle that does not enclose a singularity. This is in sharp contrast to solutions of the wave equation, which generally have less regularity.

There is an intimate connection between power series and Fourier series. If we expand a function f in a power series inside a circle of radius R, this means that

$$f(z) = \sum_{n=0}^{\infty} c_n z^n,$$

with suitably defined coefficients whose real and imaginary parts are given by,

$$c_n = a_n + ib_n.$$

Therefore:

$$f(z) = \sum_{n=0}^{\infty} [a_n r^n \cos n\theta - b_n r^n \sin n\theta]$$

$$+ i\sum_{n=1}^{\infty} [a_n r^n \sin n\theta + b_n r^n \cos n\theta],$$

which is a Fourier series for f.

Fluid Flow

Let the quantities u and v be the horizontal and vertical components of the velocity field of a steady incompressible, irrotational flow in two dimensions. The condition that the flow be incompressible is that,

$$u_x + v_y = 0,$$

and the condition that the flow be irrotational is that,

$$\nabla \times V = v_x - u_y = 0.$$

If we define the differential of a function ψ by,

$$d\psi = v\,dx - u\,dy,$$

then the incompressibility condition is the integrability condition for this differential: the resulting function is called the stream function because it is constant along flow lines.

The first derivatives of ψ are given by:

$$\psi x = v, \ \psi y = -u,$$

and the irrotationality condition implies that ψ satisfies the Laplace equation. The harmonic function ψ that is conjugate to ψ is called the velocity potential.

The Cauchy-Riemann equations imply that:

$$\varphi_x = -u, \ \varphi_y = -v.$$

Thus every analytic function corresponds to a steady

incompressible, irrotational fluid flow in the plane. The real part is the velocity potential, and the imaginary part is the stream function.

According to Maxwell's equations, an electric field (u,v) in two space dimensions that is independent of time satisfies

$$\nabla \times (u,v) = v_x - u_y = 0,$$

and

$$\nabla.(u,v) = \rho$$

where ρ is the charge density. The first Maxwell equation is the integrability condition for the differential

$$d\varphi = -u\,dx - v\,dy,$$

so the electric potential φ may be constructed to satisfy

$$\varphi_x = -u,\ \varphi_y = -v.$$

The second of Maxwell's equations then implies that

$$\varphi_{xx} + \varphi_{yy} = -\rho,$$

which is the Poisson equation.

It is important to note that the Laplace equation can be used in three-dimensional problems in electrostatics and fluid flow just as in two dimensions.

LAPLACE EQUATION IN THREE DIMENSIONS

Fundamental Solution

A fundamental solution of Laplace's equation satisfies

$$\Delta u = u_{xx} + u_{yy} + u_{zz} = -\delta(x - x', y - y', z - z'),$$

where the Dirac delta function δ denotes a unit source concentrated at the point (x', y', z'). No function has this property, but it can be thought of as a limit of functions whose integrals over space are unity, and whose support (the region where the function is non-zero) shrinks to a point. The definition of the fundamental solution thus implies that, if the Laplacian of u is integrated over any volume that encloses the source point, then

$$\iiint_V div \nabla u dV = -1.$$

the Laplace equation is unchanged under a rotation of coordinates, and hence we can expect that a fundamental solution may be obtained among solutions that only depend upon the distance r from the source point. If we choose the volume to be a ball of radius a around the source point, then Gauss' divergence theorem implies that;

$$-1 = \iiint_V div \nabla u dV = \iint_S u_r dS = 4\pi a^2 u_r(a).$$

It follows that,

$$u_r(r) = -\frac{1}{4\pi r^2},$$

on a sphere of radius r that is centered around the source point, and hence

$$u = \frac{1}{4\pi r}.$$

a similar argument shows that in two dimensions,

$$u = \frac{-\log r}{4\pi}.$$

Green's Function

A Green's function is a fundamental solution that also satisfies a suitable condition on the boundary S of a volume V. For instance, $G(x, y, z; x', y', z')$ may satisfy,

$$\nabla.\nabla G = -\delta(x - x', y - y', z - z')$$

$$G = 0 \text{ if } (x, y, z) \text{ on } S.$$

Now if u is any solution of the Poisson equation in V:

$$\nabla.\nabla u = -f,$$

and u assumes the boundary values g on S, then we may apply Green's identity, (a consequence of the divergence theorem) which states that

$$\iiint_V [G\nabla.\nabla u - u\nabla.\nabla G] dV = \iiint_V \nabla.[G\nabla u - u\nabla G] dV$$

$$= \iiint_S [Gu_n - uG_n] dS.$$

The notations u_n and G_n denote normal derivatives on *S*. In view of the conditions satisfied by *u* and *G*, this result simplifies to;

$$u(x',y',z') = \iiint_V Gf dV + \iint_S G_n g dS.$$

Thus the Green's function describes the influence at of the data *f* and *g*. For the case of the interior of a sphere of radius *a*, the Green's function may be obtained by means of a reflection the source point *P* at distance ñ from the centre of the sphere is reflected along its radial line to a point *P'* that is at a distance

$$\rho' = \frac{a^2}{\rho}.$$

Note that if *P* is inside the sphere, then *P'* will be outside the sphere.

The Green's function is then given by,

$$\frac{1}{4\pi R} - \frac{a}{4\pi\rho R'},$$

where *R* denotes the distance to the source point *P* and *R'* denotes the distance to the reflected point *P'*. A consequence of this expression for the Green's function is the Poisson integral formula. Let ρ, θ, and φ be spherical coordinates for the source point *P*. Here θ denotes the angle with the vertical axis, which is contrary to the usual American mathematical notation, but agrees with standard European and physical practice. Then the solution of the Laplace equation inside the sphere is given by,

$$u(P) = \frac{1}{4\pi} a^3 \left(1 - \frac{\rho^2}{a^2}\right) \iint \frac{g(\theta',\varphi') \sin\varphi' \, d\theta' \, d\varphi'}{(a^2 + \rho^2 - 2a\rho\cos\Theta)^{3/2}},$$

where,

$$\cos\Theta = \cos\varphi\cos\varphi' + \sin\varphi\sin\varphi'\cos(\theta - \theta').$$

A simple consequence of this formula is that if *u* is a harmonic function, then the value of *u* at the centre of the sphere is the mean value of its values on the sphere. This mean value property immediately implies that a non-constant harmonic function cannot assume its maximum value at an interior point.

Electrostatic Potential

Electrostatics is the part of physics that describes interactions between stationary charges. You are probably familiar with Coulomb's Law, the central law of electrostatics. This law says that two charged particles exert a force on each other equal to:

$$F = q_1q_2/r_{12}^2$$

The electrostatic force, F, is proportional to the product of the charges on the two particles, q_1 and q_2, and inversely proportional to the *square* of the distance separating the particles, r_{12}. Another important characteristic of a charged system is its potential energy, PE. Potential energy is created by electrostatic interactions between charge particles and is equal to:

$$PE = q_1q_2/r_{12}$$

Notice that this formula looks nearly the same as Coulomb's Law. The only difference is that potential energy is inversely proportional to the distance between charges, while the Coulomb force is inversely proportional to the *square* of the distance.

The most useful quantity for our purposes is the *electrostatic potential.* This quantity is related to PE as follows: the electrostatic potential created by a system of charges at a particular point in space, (x, y, z), is equal to the change in potential energy that occurs when a +1 ion is introduced at this point.

This definition can be made clearer with the aid of the following pictures. Imagine a molecule consisting of an electron density cloud and several positively charged nuclei.

Now suppose we want to know the electrostatic potential this molecule creates at point (x, y, z). We can obtain the potential by introducing a +1 charge at (x, y, z) and calculating the change in energy.

Index

A

Alpha Decay 150, 151
Apparatus 11, 14, 113, 135, 173, 185, 186
Assumptions 70, 71, 89, 106, 108, 109, 182
Atomic Spectra 104

B

Bjerrum Length 276, 277
Bohr Model of Atom 89
Bohr's Relationship 85
Breathing Earth 187, 188, 189
Bremsstrahlung 137, 138, 141, 142, 143, 144
Bremsstrahlung Radiation 137, 138, 144

C

Chain Reactions 197, 228
Characteristic Radiation 137, 142, 143
Characteristics 17, 25, 82, 138, 152, 155, 156, 157, 228, 230, 243, 254
Charge Interactions 272, 273, 274
Charge Separation 271, 274, 276
Charged Bodies 93, 247, 248, 249
Chlorophyll 127, 128, 129, 131,
Classification 228, 241
Conclusion 4, 10, 116, 152, 166
Conductor Ampacity 238
Conductor Size 231, 237

D

Dimensionality 203
Dissipation 270
Distributed 48, 70, 150, 250, 257, 280
Doped Semiconductors 255
Drift 14, 257

E

Electric Field 112, 230, 255, 257, 258, 263, 264, 265, 266, 267, 268, 269, 270, 271, 272, 275, 276, 279, 280, 281, 282, 283, 284, 285, 291
Electric Fields 263, 267, 276, 279
Electromagneti Spectrum 146
Electrostatic Explanation 107
Electrostatic Force 263, 272, 277, 278
Electrostatic Forces 272, 277
Emission Spectrum 95, 127, 138
Emissions 104, 114, 115, 131, 186, 187, 188, 189
Equivalence 148, 149, 196, 199, 203

F

Fast Neutron 221, 225
Floating Nuclear 222
Fluorescence 121, 123, 124, 125, 126, 127, 128, 129, 131, 132

J

Junction Breakdown 260
Junction Diode 254, 259, 260, 261
Junction Transistors 261

M

Max Planck 16, 18, 70, 71, 72, 79, 80, 95, 104, 111
Mechanisms 142
Monochrome 45, 46, 47, 49, 54, 55, 56, 57, 64, 65

N

Naphthalene 128, 130, 131, 132
Niels Bohr 97, 102, 104, 109
Nuclear Instability 147
Nuclear Structure 148

P

Phosphorescence 121, 123, 124, 126, 127, 128, 129, 130, 132,
Plutonium 155, 208, 209, 220, 221, 222, 226, 227
Preliminaries 68
Pressurised Heavy 219
Primary Coolants 226

Q

Quantization 25, 26, 27, 28, 29, 30, 31, 36, 37, 38, 39, 40, 41, 42, 43, 44, 45, 46, 47, 52, 53, 55, 56, 57, 59, 68, 70, 71, 96
Quantum Atom 101
Quinine Sulfate 126

R

Radioactive Atom 182, 183
Radioactive Decay 167, 169, 171, 172, 173, 176, 177, 178, 180, 182
Reactor Instability 152, 154
Reactor Kinetics 154, 155, 166
Relativistic Sticky 199, 200
Revisiting Balmer 88
Rutherford 90, 91, 102, 103, 104
Rydberg 86, 88, 93, 98, 109, 110, 111, 112, 113
Rydberg Constant 86, 88, 93, 98, 109, 110, 111, 112

S

Semiconductors 242, 243, 254, 255, 256, 257
Separation 19, 20, 46, 57, 106, 250, 258, 263, 264, 266, 270, 271, 274, 276
Splitability 208
Static electricity 244, 245, 263, 274, 280

T

Temperature Coefficient 157, 159, 161, 162

U

Unstable Reactors 152
Uranium Isotopes 208